MW01329001

NONSENSE OF A HIGH ORDER

The Confused World of MODERN ATHEISM

NONSENSE OF A HIGH ORDER

The Confused World of MODERN ATHEISM

Rabbi Moshe Averick

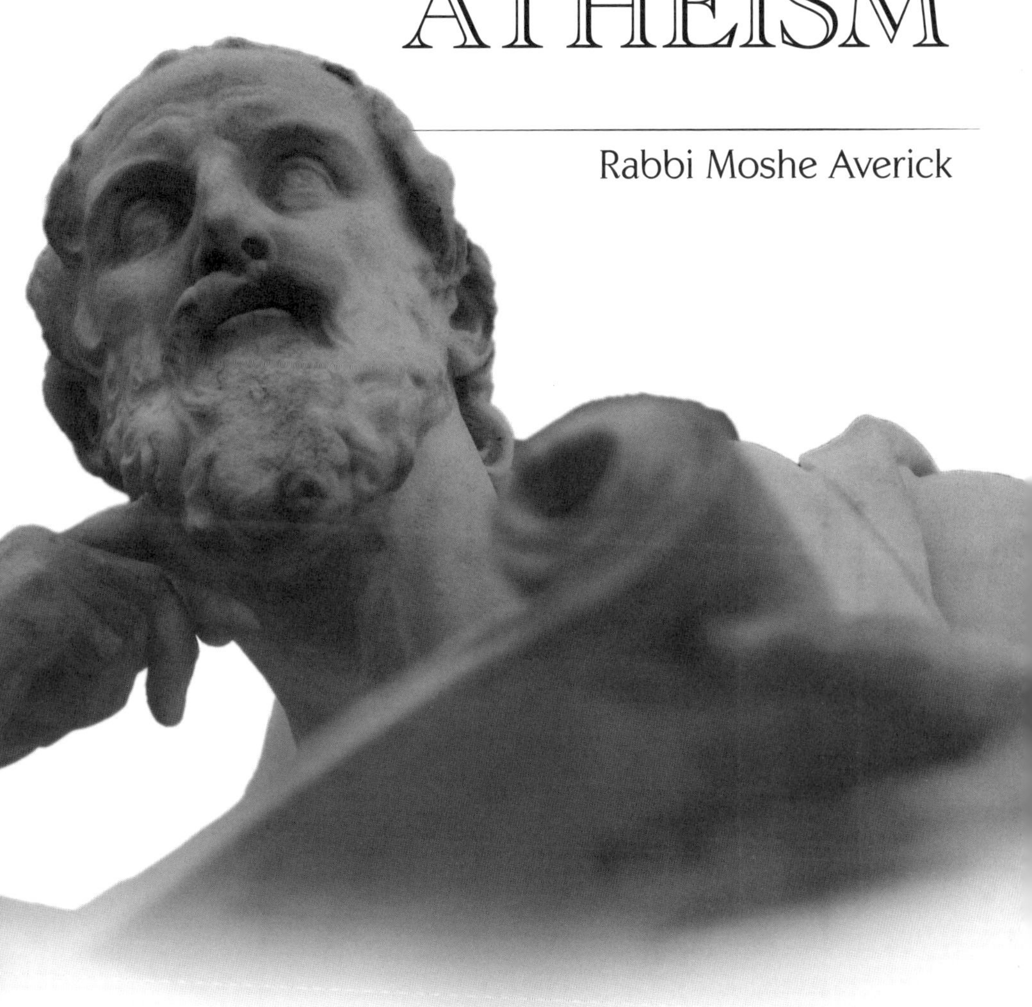

MOSAICA PRESS

Mosaica Press, Inc.

Formerly published under the title, *Nonsense of a High Order: The Confused and Illusory World of the Atheist*

Designed and typeset by Rayzel Broyde

ISBN-10: 193788757X ISBN-13: 978-1-937887-57-5

Published and distributed by:
Mosaica Press, Inc.
www.mosaicapress.com
info@mosaicapress.com

This book is lovingly dedicated
to the memory of
and for the *zchut*
and *aliya* of the *neshamah* of

Moshe ben Ze'ev Halevi

MANFRED HART

Contents

Acknowledgments

No Man Is an Island

A project such as this one is not possible without the help of many people.

First and foremost, my father and mother, Dr. Nathan Averick and Leah Averick, *a"h*.

For friendship and support through thick and thin, Michael and Mahra Hart, Rabbi Elisha and Miriam Prero, and Rabbi Yosef Kaufman and his wife, Dr. Elizabeth Kaufman.

"Coach" Miriam Ohrbach also is deserving of special mention.

Dr. Diane Medved for her encouragement and to Michael Medved who gave her the manuscript to read in the first place.

Dr. Edward Peltzer who took time out of his busy schedule to review the section on Origin of Life and Dr. Joseph Walder, CEO of Integrated DNA Technologies, Inc., who was kind enough to clarify some of the finer points of organic chemistry.

Dr. David Berlinski and Dr. Paul Nelson of the Discovery Institute who gave of their time to answer questions and critique sections of the book.

Lisa Neiberg and Mary Burt for invaluable editorial input.

Devorah Haggar who tirelessly worked on the project.

I also offer my thanks to Rabbi Yaacov Haber and his first class professional staff at Mosaica Press.

This book was also made possible in part with the assistance of NISHMA, an international organization dedicated to Torah research and education, under the direction of rabbinic scholar Rabbi Benjamin Hecht and his wife, Naomi Hecht (see Nishma.org).

Most important of all, endless gratitude to the One who "graciously endows man with wisdom and teaches insight to a frail mortal," the One "without beginning or end," my Rock and Redeemer, the One God of all mankind.

Section 1

Fundamentals

Chapter 1

Introduction to Modern Atheism 101

I remember quite distinctly the afternoon I walked into a book store in Chicago and found myself encircled by a phalanx of best-selling atheist manifestos displayed in strategic locations all over the store; books such as Richard Dawkins' *The God Delusion,* Christopher Hitchens' *God Is Not Great: How Religion Poisons Everything,* and other titles, like *The End of Faith, God: The Failed Hypothesis*, and *Why Evolution Is True*. My reaction to seeing these books may surprise you: I was thrilled!

I imagine that the sensation was similar to that which a WWI pilot (an era when there still was a sense of chivalry between warring opponents) in his Sopwith Camel biplane might have felt as he engaged a worthy opponent in a dogfight — relishing the opportunity to do battle with a skillful foe, yet brimming with confidence that he would emerge victorious. There are few things I enjoy more than a vigorous, honest battle of intellect, pitching worldview versus worldview and idea versus idea. I soon became acclimated to my surroundings and thoroughly enjoyed the simple pleasure of sitting in a comfortable chair, drinking coffee, and reading.

I have taught Jewish theology and Judaic studies for nearly thirty years and I lost track long ago of the countless hours I have spent (with thousands of students of all ages) teaching, discussing, arguing, and debating the existence of God, spirituality, the ultimate purpose of our lives, etc. For several months after that day in the store, I spent day after day, hour after hour, wading through the combined onslaught of the twenty-first century's most popular atheistic ideologues. However, my initial excitement eventually dissipated and finally faded away.

I had hoped to find in these manifestos at least a little bit of cutting-edge intellectual searching and honesty. I was sorely disappointed. While Richard Dawkins' pronouncement in *The God Delusion,* that "the God of the Old Testament is...petty, unjust, vindictive, bloodthirsty, misogynistic, homophobic, racist, infanticidal, genocidal, etc.,"[1] may, for some, have an irreverently bold and strident ring, it is more the proclamation of a stance than it is the outcome of intellectual inquiry. Can an intellectually honest and open-minded person ignore the fact that this same Deity commands the Israelites not to take revenge or bear a grudge, to view the use of inaccurate weights in business as an abomination, to view all human beings as created in the image of God, to open our hands wide to those in poverty, not to oppress the stranger who lives among us, to leave a portion of every field unharvested for the poor, never go to war against an enemy without first offering peace, to "love your neighbor as yourself," and that "justice, justice shall your pursue"?

Jewish Scripture is the single most influential piece of literature in the history of mankind. Nothing else even comes close. Is Dawkins *obligated* to agree with the following passage written by distinguished historian Paul Johnson in his monumental historical treatise, *A History of the Jews*?

> To them [the Jews and their Scripture], we owe the idea of equality before the law, both Divine and human; of the sanctity of life and the dignity of the human person; of the individual conscience and so of personal redemption...of peace as an abstract ideal and love as the foundation of justice, and many other items which constitute the moral furniture of the human mind.[2]

Perhaps not; but even if Dawkins has decided to completely reject the biblical worldview, it stands to reason that at least a *slightly* more nuanced evaluation would have been in order. It's clear to me that the chain of venomous one-dimensional invective cited above offers us much more insight into the inner workings of the soul of Richard Dawkins than it offers us any meaningful insight into understanding the biblical narrative or the concept of the One God who is at the center of it all.

While I have never been particularly impressed by the intellectual firepower brought to bear by skeptics, atheists, and freethinkers in their attacks on belief in God, I have striven to respond seriously to their expressed views. I wish the same could be said for our new breed of militant skeptics. Bombastic titles like *God Is Not Great: How Religion Poisons Everything*, (everything?!); rantings about people who profess religious faith: "When their beliefs are...common we call them "religious," otherwise they are...called "mad," "psychotic," or "delusional";[3] and the rehashing of philosophical parlor tricks like, "Can God create a stone that is too heavy for him to lift?"[4] do not, in my opinion, contribute to the expansion of our intellectual or spiritual horizons.

It was clear to me that the gauntlet had been thrown down. I resolved on the spot to write a book in response — a book whose essential purpose would be to demonstrate without question that it is the theist who wields the decisive advantage on the intellectual battlefield. As a fitting introduction to our subject matter, consider making the acquaintance of one of the most charismatic of the high clergy in the modern atheist hierarchy, the late Christopher Hitchens.

An Atheist "Sees the Light"

Hitchens (1949–2011), who was educated at Oxford, was a highly accomplished author and journalist. He wrote books on Thomas Jefferson, Thomas Paine, Henry Kissinger, and Bill Clinton, among others. His columns and articles appeared in prestigious publications around the world and he was regularly interviewed on radio and television. In 2007, his atheistic magnum opus was published under the title *God Is Not Great: How Religion Poisons Everything.* This work was followed swiftly by a compilation entitled *The Portable Atheist: Essential Readings for the Non-Believer.* In a debate that took place shortly after the publication of *God Is*

Not Great, Hitchens recounted to the audience how even as a young boy he "saw the light" of atheism:

> I was nine when I thought I saw through it, when my biology teacher told me that God was so good as to have made vegetation green because it was the color most restful to our eyes. And I thought, Mrs. Watts, this is nonsense...I just knew she'd got everything all wrong.[5]

As Hitchens grew older he received a fine education and acquired an admirable command of the English language. Nevertheless, I suspect that a careful analysis of his views on such weighty issues as racism, morality, and democracy — considered through the prism of his staunch atheistic worldview — might in fact leave an honest thinker wondering if his intellectual progress in this particular area didn't stall at the level of a nine-year-old boy daydreaming during Mrs. Watts' science class.

In a September 2007 lecture at Sewanee University in Tennessee, peculiarly entitled, "The Moral Necessity of Atheism," Hitchens expressed his loathing of the "primitive" concept of racism. He resolutely advanced the argument that atheism, with its implicit notion that human beings merely represent another evolutionary branch of the animal kingdom, deals a fatal intellectual blow to racism as a concept:

> Through the DNA we find that in some sense, some of Genesis is vindicated. We are in a way part of an animal creation and we share part of their material in our own make up. I, as a mammal, never kind of doubted that I had this relationship with ground worms and other creatures. It does make short work of racism. It means racism is no longer something we have to argue with...we may be some distance from being able to completely pronounce its utter defeat, but it's over as an argument...racism is a primitive, stupid construct made out of literally nothing.[6]

In other words, if we would accept the premise that our existence here on earth is the result of Darwinian evolution; if we would acknowledge, as non-believing paleontologist Stephen J. Gould has put it, that "we are here because one odd group of fishes had a peculiar fin anatomy that could transform into legs";[7] if we would reflect on our shared relationship to ground worms, our response would be to join hands and triumphantly break into a chorus of *We Shall Overcome*.[*] Really? To be fair, if the atheist/Darwinian view is accurate, we *are* all brothers in the sense that we are equally related to ground worms. Astonishingly though, Hitchens — perhaps blinded by the brilliance of his third-grade epiphany — did not acknowledge what seems to be the ultimate emptiness of such a statement.

A shared biological relationship with ground worms does not make all of mankind brothers in the sense that we are all equally *valuable*. It makes us brothers in the sense that we are equally *void of all significance*. A ground worm is insignificant. There is nothing ennobling or inspiring in one's being related or equated to a ground worm. Thus, the species *Homo sapiens* is also insignificant. In the atheistic worldview, both are nothing more than infinitesimally small specks of "dust in the wind"; random, meaningless collections of molecules and chemicals spinning in space. But why should I bore the reader with *my* rendition of the atheist view of reality? There is no reason they should not be allowed to speak for themselves. Dr. Peter Walker, a space physicist at Rice University informs us that:

> [Humans] are carbon based bags of mostly water on a speck of iron-silicate dust revolving around a boring dwarf star in a minor galaxy in an underpopulated local group of galaxies in an unfashionable suburb of a supercluster.[8]

Similarly, the eminent astrophysicist Sir Arthur Eddington proclaims, "We are bits of stellar matter that got cold by accident, bits of a star gone wrong."[9]

The late astronomer, Carl Sagan, leaves very little to the imagination as far as his viewpoint of mankind's place in the grand scheme of things: "The very scale of the universe...speaks to us of the inconsequentiality of human events in the cosmic context."[10] And finally science educator, Bill Nye[**]:

---

* The inspirational and iconic anthem of the civil rights movement of the 60s and 70s.

** Known on his popular PBS television show as "Bill Nye the Science Guy."

> I am just another speck of sand and the earth really in the cosmic scheme of things is another speck and the sun, an unremarkable star...is another speck, and the galaxy is a speck. I'm a speck on a speck, among other specks, among still other specks![11]

In a Godless, material universe, all life on earth drowns in an ocean of insignificance in relation to the countless billions of galaxies that surround us.

I fervently agree that racism is "stupid." Other than that, Hitchens and I have nothing in common regarding our views on racism. For a theist such as myself, racism is "stupid" because all human beings are created in the image of God; all stand equal before their infinitely powerful Creator. The intrinsic value of a human being derives from his relationship to God and is not predicated on his physical and mental abilities and certainly not the color of his skin. For Hitchens, however, racism is "stupid" not because of some noble notion of the exalted *brotherhood* of all men, but because of the absolute *insignificance* of all men, no matter what their race or color. The distinguished evolutionary biologist George Gaylord Simpson has informed us that "Man is the result of a purposeless [evolutionary]...process."[12] H. L. Mencken put it a little more bluntly: Man is a "sick fly" spinning around in space on a "dizzy ride" to nowhere.[13]

In a purely physical, material universe, man's life is ultimately insignificant and his death is insignificant. His words and thoughts are insignificant and his endless collection of ideologies is insignificant. Not only is the skin color and racial type of a "sick fly revolving around a boring dwarf star" not worth a moment's thought, *nothing* about his existence is worth a moment's thought. Despite the ultimate non-value of human existence, in an honest analysis of his view of reality, Hitchens stubbornly persists by expressing not only his distaste for racism, but his enthusiastic support of democratic principles. In the same lecture at Sewanee University, Hitchens proclaimed his appreciation of American democracy:

> The American Revolution is the only one still standing...the only one that has any merit or virtue left in it, and I think this confers

> upon us a certain responsibility...[It] should be a great deal better appreciated than it is, and a great deal more cherished, and a great deal more firmly upheld at home as well.[14]

Though an admirable tribute, Hitchens seems to ignore a very *inconvenient* truth. If racism is "stupid" for the reasons that he advances, then one could conclude that the ideology of democracy must be considered as even "stupider." American Democracy is built on the premise (as stated in the Declaration of Independence) that "we hold these truths to be self-evident, that all men are created equal, that all men are endowed by their Creator with certain unalienable rights, that among these are life, liberty, and the pursuit of happiness." This makes perfect sense to me because I believe that all men are created in the image of God and all stand equal before their infinite Creator.

An intellectually honest atheist, however, would cringe at these words. He would bring a lawsuit claiming that reading the Declaration of Independence in a public school is a violation of the separation clause of the Constitution. Not only are men not created equal, they are not created at all. We are only here because, "one odd group of fishes had a peculiar fin anatomy that could transform into legs." Not only are men not endowed with unalienable rights, there is no Creator to endow them with any inherent rights whatsoever. In Darwinian terms, as described above, we are a kind of glorified tuna fish. What inherent rights does a tuna have? (The right to be picked by Star Kist?)

In his introduction to *The Portable Atheist*, Hitchens declares, "I am writing these words on July 4, 2007, the anniversary of the proclamation of the world's first secular republic." It is unmistakably clear from the Declaration of Independence itself that this statement is an outright falsehood. No one has ever stated the obvious truth about American Democracy with more clear and penetrating incisiveness than the early twentieth century writer and thinker, G. K. Chesterton:

> The Declaration of Independence dogmatically bases all rights on the fact that God created all men equal; and it is right, for if they were not created equal, **they were certainly evolved unequal**.

> There is no basis for democracy except in a dogma about the Divine origin of man.[15]

For the non-theist, the notion that all men are equal — albeit a comforting fiction for some — has no basis in reality. In what way are they equal? Some are brilliantly intelligent and some are amazingly stupid. Some are highly competent and talented, some are completely inept. Some are robust and powerful, and some are sickly, crippled, and weak. Some are clearly born to lead and some seem born to follow. Most significantly, for those who take Social Darwinism seriously, some are fit to survive and some are *not* so fit to survive.

In my opinion, one of the more egregious intellectual blunders of atheists like Hitchens is to espouse noble ideals (like democracy and equality) that only make sense if a transcendent God/Creator exists, and when no one is looking to quietly drop him out of the picture. They then hope nobody notices that removing God from the equation effectively destroys any possible rational foundation for the very ideal they are promoting. If Christopher Hitchens wanted to *subjectively* assign a higher value to the ground worm-related, upright-walking primate called *Homo sapiens* or to some particular human-manufactured ideology, then by all means. However, he should have displayed enough intellectual courage and integrity to admit that from the viewpoint of the atheist/materialist, these are artificial constructs "made out of literally nothing" that have no objective reality. Other atheistic thinkers and philosophers were a little more candid on this subject.

In a letter to Marie Bonaparte in 1937, Sigmund Freud wrote, "The moment a man questions the meaning and value of life, he is sick, since objectively neither has any existence."[16] Freud's conclusions are echoed by another well-known non-believer, the Nobel Prize-winning physicist Dr. Steven Weinberg: "The more we know of the cosmos, the more meaningless it appears."[17] Professor Will Provine of Cornell University explains that atheism inescapably leads to the conclusion that "there is no hope whatsoever of there being any deep meaning in human life...you're here today and you're gone tomorrow and that's all there is to it."[17]

Freud, Weinberg, Provine and others state this simple, obvious truth that follows from the atheistic worldview. In fact, Hitchens himself admits to the problem. Later on in the same lecture he responds to a question from the audience as follows: "The question for me would rather be, this being the case, [that there is no purposeful creation] then why care, why do I bother? That's a very good question. **It also doesn't have a conclusive answer**."[19] Let's step back and think about this for a moment: Religion poisons *everything*, atheism is "necessary for morality," but inasmuch as there is no purposeful creation there is really no clear reason to care or bother? It seems to me that before embarking on his whirlwind crusade for non-belief, Hitchens should have cleared up that question first. If I were a comedian and wanted to parody Hitchens' lecture at Sewanee University, it would have gone something like this:

> *My fellow ground worms,*
>
> *It is important for you to know that religion poisons everything! Imagine how beautiful life would be if only we would stop trying to treat our fellow man like he was created in the image of God, stop treating him as if the Creator endowed him with unalienable rights, stop pretending that all men stand equal before their Creator, and start treating him like a purposeless carbon-based bag of water revolving around a boring dwarf star, like bits of stellar matter gone wrong, like a sick fly, like the ground worm that he is...*
>
> *Remember, it is religion that poisons everything! The only thing religious people do is to go around killing each other in the name of God. I ask you honestly, does any rational, logical, skeptical, atheistic, scientific minded person really think it's necessary to believe in God if you want to go around killing people? Of course not! Don't let those fanatics brainwash you. Stalin and Pol-Pot* murdered* **millions** *and let me remind you that they were atheists. Anything those religious people do, we can do at least as well, if not much better.*
>
> *Let's be frank; as an avowed non-theist, I have to admit that Thomas*

* Joseph Stalin, tyrannical dictator of the Soviet Union from 1924–1953; Pol-Pot, Prime Minister of Kampuchea (Communist Cambodia) from 1976–1979.

> *Paine's idea that all men are* **created** *equal is a "stupid" construct "made out of literally nothing." By the same token, Adolf Hitler's racist idea that the Aryan race is superior to all other races is also a "stupid" construct "made out of literally nothing." There is, however, an important difference between the two — Thomas Paine did not have that silly moustache.*
>
> *In the final analysis I really don't know what difference it makes anyway, since I have no conclusive reason to care one way or the other. I guess inconclusiveness is better than nothing; in my case it certainly pays the bills. However, please keep in mind that prominent non-theists such as Dr. Sigmund Freud, Dr. Steven Weinberg, Dr. Will Provine, etc., have made it clear that there is no objective purpose or value to human life, and the universe is meaningless and pointless. This of course means there is no real point in me speaking to you, or for you to listen to me for that matter...which makes me wonder...why exactly do I keep speaking anyhow?... But even more important, why do you keep listening? The main thing to remember is this: All these things are necessary for morality!*

How it is possible on the one hand to assert that one's worldview implicitly condemns racism, promotes democracy and the equality of all men, while at the same time admitting that the very same worldview offers one no conclusive reason to care or bother about anything, is rather difficult for me to understand. There seems to be a profound disconnect operating here. I contend that these types of disconnects, contradictions, and inconsistencies typify much of atheistic thought, as I will attempt to show throughout the book. Non-theists/atheists are certainly not the only ones who are inconsistent in presenting their ideas or who disconnect from certain truths and realities in order to maintain a belief in a particular worldview; in all fairness they would most certainly accuse me of doing the same thing. In the final analysis, though, it is up to the individual to carefully weigh the evidence and make a decision. However, before proceeding to the heart of the matter regarding the existence of God, it is imperative to set out some basic philosophical underpinnings and conceptual paradigms for our discussion. I call these "The Ground Rules."

End Notes

1 Richard Dawkins, *The God Delusion* (New York: Houghton Mifflin Co., First Mariner Edition, 2008), p. 52.
2 Paul Johnson, *A History of the Jews* (New York: Harper and Rowe, 1988), p. 585.
3 Sam Harris, *The End of Faith: Religion, Terror, and the Future of Reason* (New York: W.W. Norton and Company, 2004), p. 72.
4 Victor Stenger, *God: The Failed Hypothesis: How Science Shows that God Does Not Exist* (Amherst, New York: Prometheus Books, 2007), p. 33.
5 Christopher Hitchens Debates Al Sharpton: New York Public, YouTube.com.
6 "The Moral Necessity of Atheism," a lecture by Christopher Hitchens at Sewanee University (2004), YouTube.com
7 Joan Konner, *The Atheist's Bible* (New York: Harper Collins, 2007), p. 2.
8 Jack Huberman. *The Quotable Atheist* (New York: Nation, 2007), p. 313.
9 Ibid., p. 98.
10 Carl Sagan, *Broca's Brain: Reflections on the Romance of Science* (Toronto: Ballantine Books, 1980), p. 341.
11 Bill Nye speaking at the 2010 American Humanist Association Conference, YouTube.com
12 G. Gaylord Simpson, *The Meaning of Evolution*, cited in David Oderberg, *Real Essentialism* (New York: Routledge, 2007), p. 241.
13 http://en.wikiquote.org/wiki/H._L._Mencken
14 See note 6 above.
15 G. K. Chesterton, *What I Saw in America* (1922), Chapter 19, http://chesterton.org/acs/quotes.htm (under *Government and Politics*).
16 *Letters of Sigmund Freud*, edited by Ernst Freud (New York: Basic Books, 1960), p. 436.
17 From *Dreams of a Final Theory*, as quoted by Dr. Stuart Kauffman, http://www.edge.org/3rd_culture/kauffman06/kauffman06_index.html
18 From an interview in the film *Expelled: No Intelligence Allowed*, http://www.youtube.com/watch?v=vuVSIG265b4&feature=related
19 See note 6 above.

Chapter 2

The Ground Rules: Guidelines for Discussing the Existence of God

Please Note: It is important that the reader not skip over this chapter

Ground Rule #1: Seeking the Truth

This book is about the crucial decision human beings face regarding the existence of God. It is without a doubt the single most important decision we ever make. The entire meaning, purpose, and direction of our own lives, the way we relate to our families and loved ones, and the entirety of humanity, hangs on this decision.

It is my opinion that the reality of the Creator, or the existence of God, is a truth that is quite accessible. Although the ideas that I will present may require careful analysis and contemplation, they are not particularly complicated or difficult. Most can be understood by an intelligent and inquisitive high school senior. Not everybody has been trained to analyze and evaluate these types of issues. This is not necessarily a reflection of an individual's native intelligence; proper analysis of an abstract or philosophical concept is a skill he or she may never have acquired. Regarding this, I hope I have presented my ideas in a clear enough fashion that even newcomers will be able to follow.

There is a second challenging factor that is much more serious and cuts to the core of the essential nature of a human being. Quite simply, it is the *desire for truth*. If people have not assigned "truth" a top slot on their list of priorities, it does not matter what is said to them and what evidence is presented. The following example will help illustrate this universal behavioral tendency that human beings frequently exhibit when confronted with uncomfortable facts or ideas.

On March 13, 1964, a young woman named Kitty Genovese was stabbed to death in front of her apartment building in the Kew Gardens section of Queens, NYC. What made this crime so notorious and famous was the claim that thirty-eight witnesses had watched from the large apartment complex as she was murdered and not one of them called the police. That issue is not our point here. The murderer, Winston Moseley, was eventually caught, tried, and incarcerated. Moseley was once interviewed in prison. During the interview he spoke freely about the fact that he stabbed this young woman to death, how much he regretted what he had done, and how much he had changed during his years in prison.

A few moments later in this documentary film, Moseley's mother is interviewed and she emphatically declared that she "knows" that her son never killed anyone. She repeated this several times. The juxtaposition of these two scenes — of Moseley himself describing how he stabbed a woman to death and then of his mother declaring that she "knows" that her son is innocent — was jarring to say the least. Her unequivocal and convincing manner would surely have planted seeds of doubt about his guilt in the mind of anyone who had not previously seen the interview with Moseley himself.

"A Man Hears What He Wants to Hear, and Disregards the Rest"

(from "The Boxer" — Simon and Garfunkel)

The conclusion is obvious. If a person does not want to believe even the most obvious fact, if there is an agenda that is more important than the truth, it is quite possible that there is nothing that will convince that person. This woman's psychological and emotional need to believe her son to be innocent was much greater than her desire for the truth. Consequently, she "knows" that her son did not commit murder.

Although it is easy for most people to understand why this woman was prepared to deny reality to fit her own agenda, the point is that we are all perfectly capable of doing the same thing. Actually, to say we are *capable* of doing it is very misleading. It is much more than that: The struggle to seek the truth and let the truth determine our agenda, rather than use our personal agenda to determine what we will accept as truth, is the essential battle we must be prepared to fight if we wish to be fully human. The alternative to finding the truth and living with the truth is living with lies, falsehoods, and delusions.

The late Rabbi Noah Weinberg* once noted that "every human being exhibits a Nobel Prize level of genius and creativity when it comes to one particular area: *rationalization*."[1] The most dangerous enemy of truth is not a lie. The real enemy of truth is when we ourselves *choose* a comfortable lie of our own creation over the truth.

A striking dramatization of this basic human struggle was presented in the 1999 cult-film classic, *The Matrix*: At some future time, technologically advanced machines have taken over the world. Human beings are kept in a sleep state where their brains are hooked up to a massively complex computer program called "The Matrix." The computer program gives these comatose people the illusion that they are actually living a real life as we know it. A few humans have escaped and are fighting the machines. One of these escapees makes a deal with the machines: in return for betraying the rest of his group, he will be reconnected to "The Matrix" and in this *illusory dream state* will be someone "very important." He chooses a comfortable dream over actual living.

We all face the same dilemma. Am I prepared to make the sacrifices necessary to live with reality or, like the traitor in that film, will I choose a comfortable fantasy and illusion instead? How badly do I want the truth?

* (1930-2009) Founder and Dean of the Aish HaTorah Rabbinical Seminary, Jerusalem.

Would I be prepared to live in poverty for the truth? Would I be prepared to take a *10% cut in income* for the truth? If the price for finding the truth was living on bread and water for the rest of my life, would I pay that price? How much am I willing to sacrifice, how much comfort am I willing to forgo, to find the truth? The danger every individual faces is ending up like Winston Moseley's mother, or the traitor in the film, denying an obvious reality because it threatens his or her comfortable way of life, ideology, or agenda.

Ground Rule #2: Defining Fanaticism

Whenever religion, God, or the concept of truth is being discussed, invariably someone is going to start throwing around the term *fanatic*.

I have been labeled by some as a fanatic simply as a result of identifying as an Orthodox Jew. My experience has been that most people have never bothered to think about what this word actually means or implies and will stare dumbfounded if asked to state a meaningful and useful definition of the word. Usually they offer something like, "religious people who believe they have the truth are fanatics," or "people who think they are right and everyone else is wrong are fanatics," or "people who will kill for what they believe in are fanatics." Of course, none of these are definitions; furthermore, when followed to their logical ends, they lead to quite foolish conclusions.

- *Religious people who think they have the truth are fanatics*. According to this understanding of the concept, what makes them fanatics? Is it (a) because they think they have the truth, or (b) because the assumption is that their beliefs are false? If it is (a), then that would mean *anyone* who thinks he knows the truth is a fanatic. That would include people who believe the earth is round and biologists who believe in Darwinian evolution. If it is (b), then it has nothing to do with the fact that they are religious per se, but because anyone who believes ideas that are false is a fanatic. This is quite unsatisfying because you have now begged the question: What is the truth? It goes even further; is this to say that every time someone holds a mistaken idea he is considered to be in the category of a fanatic? Scholars constantly disagree and revise theories. Does this mean that if one of their theories turns out to be mistaken (i.e., false) they were, until now, a fanatic?

- *People who think they are right and everyone else who disagrees is wrong are fanatics.* This, of course, is ridiculous. It would mean that anyone who is convinced he is right about a particular issue should be labeled as a fanatic. There is no escaping from the fact that if a person thinks he is clearly right on a particular subject, there is an implicit declaration that anyone who disagrees is wrong.
- *People who kill for what they believe in are fanatics.* That would mean that the Allied soldiers who fought the Nazis in WWII were fanatics. It would mean that all policemen who shoot at armed robbers are fanatics. They were and are ready to kill and die for what they believe in.

We clearly have to do much better than this if we are going to understand what fanaticism really is and the significance of the concept in our discussion. The clearest and most meaningful definition of a fanatic I ever heard was the one proposed by Rabbi Noah Weinberg:

> Fanaticism has nothing to do with what you believe or how passionately you believe it. The definition of a fanatic is someone who believes in a particular idea or doctrine no matter what it may be, and says: "This is what I believe, it's the truth, and don't disturb my comfortable beliefs and ideas with your questions, observations, and certainly not with facts. I'm right, I'm comfortable, and I am determined not to think about it anymore."[2]

What distinguishes the passionate believer in a cause or ideology from the fanatic is not what he believes but whether or not he is open to questions, to consider another point of view, or to be theoretically ready to reconsider his position. The fanatic "discusses" or "argues" only to justify his beliefs and agenda, but never to discover truth. In short, fanaticism is not a function of *what* one believes; it is a function of how — on an internal level — one emotionally and intellectually *relates* to those beliefs. The fanatic is someone who has *shut off his or her mind* and is unwilling to consider anything that presents a challenge to his dearly held belief system.

A Litmus Test for Fanaticism

We now have a meaningful definition of fanaticism. It becomes clear then that there is *no ideology* that is free from fanatics. The fact that a person advocates his or her position with great passion and self-sacrifice has nothing to do with fanaticism. A person can present his position very calmly, quietly, politely, and still be a total fanatic. Winston Moseley's mother is a fanatic when she denies her son is a murderer; there is nothing anyone could say, no matter how reasonably stated, that would change her mind. If someone were to show her the interview of her son admitting to murder, she would simply claim that he was coerced or threatened with torture.

This means that determining who is really guilty of fanaticism is not always such an easy task. However, there is at least one (almost) surefire *indicator* that someone is a fanatic. When an individual expresses indignation at the suggestion that there are adherents of his or her *own* ideology who might be fanatics, or denies the possibility of such, it is almost always an indication of the fanatical nature of that particular individual. For example, if you meet a feminist ideologue who denies that there could be such thing as a fanatical feminist or who bristles at the suggestion that such a thing is possible, it is highly likely that she herself is a fanatic. Only someone who has shut off their own mind could be so blind as to deny the possibility of fanaticism in his own particular ideological community. The same applies to religious believers, freethinkers, homosexual activists, global warming activists, pro-abortion activists, scientists, conservatives, liberals, socialists, capitalists, and atheists, to name a few.

Ground Rule #3: It Is Critical to Clarify How We Know and Decide that Something Is True

The philosophical term for this idea is *epistemology*. How do we know what we know? What does it mean to "prove" something? What does it really mean when we say we "know" something is true? In light of the fact that we are talking about knowledge of the existence of God, it is quite important to spend a little bit of time on this subject.

A fundamental decision that every human being must make is whether or not it is possible to bridge the gap between oneself and reality. In other words, can we actually know and discern the world around us? When confronted with uncomfortable new ideas about God, religion, and morality, I have often heard people respond, "Well, maybe we don't really *know* anything. Maybe we are just dreaming this reality," (or some variation on that theme).

How we know the world around us is real, or *how* we are sure of our own reality, can be a fascinating subject to discuss. Quite a bit has been written on those questions and a full discussion is clearly beyond the scope of this book. However, there is no escaping the fact that ultimately it boils down to how you choose to answer the following question: Do you trust your senses, your mind, and your brain, to give you accurate data with which to interface with reality? Please answer yes or no. There are no other options. If your answer is "no, I do not trust that my mind and senses can convey accurate information with which I can form a coherent picture of reality," discovering the truth about *anything* is going to be a rather difficult task. On the other hand, for our purposes, if you answered "yes," please continue reading and it will not be necessary to bring up the subject again. Answering "yes" does not necessarily mean that it is possible to know and understand everything and anything; it just precludes using "we can't really *know anything*" as an excuse to avoid dealing with uncomfortable ideas.

Once the decision is made that our ability to understand reality and the world around us is one that is meaningful and significant, it becomes relatively easy to formulate at least a working definition of what it means to "know" and/or to "prove" something: *knowing* means to know beyond a reasonable doubt, while *proving* something means to present evidence that demonstrates the proposition to be true beyond a reasonable doubt.

It is not possible to know anything or to demonstrate anything to a higher level of clarity than "beyond a reasonable doubt." Absolute certainty with no possibility of error is not our goal because it is unattainable. There are of course many different lower levels of clarity where the odds might be in your favor but there still is reasonable doubt. I would suggest that both what we call "mathematical" proof and "scientific" proof fall into the same category. It is crucial to understand why even in science and math the highest level of clarity achievable is "beyond a reasonable doubt."

> We all recognize the typical courtroom drama where the defense attorney in a criminal trial approaches a witness and says, "Mr. Jones, we have heard your testimony earlier in this trial...isn't it *possible* you made a mistake? Mr. Jones, please answer the question, isn't it *possible* you made a mistake?" Mr. Jones hesitates and mumbles, "Well it's possible, but..." and the defense attorney immediately interrupts and says, "That will be all Mr. Jones, you may step down." Then, while the witness sits there sputtering, the judge ends it by turning to Mr. Jones and saying, "The witness is excused." The defense attorney slyly smirks because he has effectively undermined the credibility of the witness, the prosecuting attorney grimaces at the underhanded tactic of his opponent, and the witness is left gaping foolishly.

In fact, there is an easy remedy for the prosecuting attorney. Since what is required in a criminal trial is proof beyond a *reasonable* doubt, he simply has to ask the witness the following: "Mr. Jones, you said it's *possible* you made a mistake, however, do you have any *reasonable doubt* that you made a mistake? Is there any *reasonable doubt* that your testimony is not accurate?" Of course, the answer to that question would be an emphatic "no!"

My point is very simple. No matter what propositions you are dealing with, whether theories in the fields of statistics, math or science, guilt or innocence in criminal trials, or the issue of the existence of God, one can always raise *unreasonable doubt*. I repeat, no matter what issue we are discussing you can *always raise unreasonable doubt*. Ask yourself, isn't it *possible* the roof will cave in within the next thirty seconds? If it's *possible* then why don't we run out immediately? The answer is simple; it may be possible, but it is not reasonable. The exact same principle applies to scientific experiments:

Consider a hypothetical experimental trial of a new medication with two groups of 1,000 cancer patients. In Group 1, the group that receives the new medication, 999 are cured. In Group 2, which does not receive the new medication, nobody is cured. I go to another hospital and repeat the experiment with the same results. Isn't it *possible* there is some explanation other than the medicine to explain why the patients are healthier? Yes, it's *possible*, but at some point in the experiment it ceases to be *reasonable*.

Even 2+2=4 can be doubted (*unreasonably*, of course). Many people are incredulous when they hear that statement and ask, "How can anyone doubt that 2+2=4? Isn't it mathematically inescapable?"

In response to this, I have presented my students with the following scenario: Einstein's Theory of Relativity is scientifically accepted across the board. Part of that theory is that time is relative. As we move at a faster velocity or in the presence of a large gravitational field, time actually slows down relative to people who are moving slower, or relative to people who are in the presence of a smaller gravitational field. This scientifically proven fact is perhaps one of the most counter-intuitive realities that we simply have to accept as being true. Before Einstein published the theory and it was proven conclusively, you would have thought that the idea of time being relative was totally insane.

So now I ask you: Isn't it *possible* that there is a mathematician or physicist somewhere right now who is coming up with a new scientific theory that will show us that in the same way we were all mistaken about the true nature of time, we have all been mistaken about the true nature of 2+2=4? Isn't it *possible?* I have now succeeded in raising unreasonable doubt in my students' minds; of course, they have no choice but to answer "yes."

We see then that it is *possible* — but totally *unreasonable* — to doubt that 2+2=4. It is crucial to realize that when considering any type of scientific or philosophical issue, I do not have to consider unreasonable doubt. I do not have to consider *every* theoretical possibility, only *reasonable* possibilities.

Another example: Two men go into a sealed bank vault with only one door. Suddenly, a shot rings out. One man is found holding a smoking gun and the other is dead on the floor from a gunshot wound. At the trial, the defense attorney says to the jury, "Isn't it possible that a crew member from the Starship Enterprise used his futuristic technology to beam himself back in time and down into the bank vault, shoot the man, stick the gun in my client's hand, and then beam back up? Isn't it possible?"

The answer of course is that we don't care one bit if it's *possible*. We only care if it's *reasonable*. I have heard many unreasonable doubts and arguments raised when it comes to the issue of the existence of God. Ridiculous possibilities and permutations of possibilities that no one would dream of bringing up for anything else, all of a sudden become standard fare for this issue. We do not need to demonstrate or know "mathematically" or "scientifically" that God exists. We just need to demonstrate it or know it *reasonably*. Lest anyone doubt the power and significance of knowing something reasonably, we stake our lives (every day) on reasonable propositions.

When you cross the street, you calculate how much time you have until the truck coming in your direction reaches you. How do you know for sure that your brain is calculating the distance and speed properly? Maybe the truck is in reality much closer. It certainly is *possible*. Every single day, millions of people enter aircraft and fly all over the world. To get into an airplane with a pilot who does not know how to fly is to risk almost certain death. How many people have "proof" that the pilot actually is qualified? How many people have gone up to the cockpit and investigated who the pilots are and if they are certified? It simply is a highly reasonable proposition that the uniformed men in the cockpit are really pilots. Conversely, it is a highly unreasonable proposition that anyone could get into the cockpit of the airplane who was not a qualified pilot.

If you are ready to risk your life without a moment's hesitation on a very reasonable proposition (remember this the next time you step into an elevator), please don't engage in sophistry by pulling Hume and Kant* out of your hat when it comes to the existence of God. There are any number of philosophical ideas that seem startling and fascinating inside

---

* David Hume and Immanuel Kant: Two highly influential eighteenth century philosophers who wrote seminal works on epistemology and metaphysics.

the university lecture hall, but once outside have little or no connection to our everyday reality.

Ground Rule #4: There Is a Difference between Belief in God and Belief in a Particular Religion

In November 2006, there was an e-mail debate about the existence of God between conservative radio talk-show host and author Dennis Prager and atheist author Sam Harris. The debate was entitled "Why Are Atheists So Angry?" In Sam Harris' opening email, he wrote the following: "Incompatible beliefs about this God [i.e., the God of monotheism], long ago shattered our world into separate moral communities — Christians, Muslims, Jews, etc. — and these divisions remain a continuous source of human violence."[3]

Leaving aside the issue of religion being a source of strife and violence (it will be dealt with elsewhere in the book), Sam Harris presents an inaccurate description of what separates Jews, Muslims, and Christians. It is not "incompatible beliefs about this God" that creates the separation; it is incompatible beliefs of what this God *has revealed to mankind*, or perhaps more specifically, incompatible beliefs as to *how this God wants us to relate to those outside of the particular faith community*. The beliefs about God himself however are quite compatible.

The basic theological understanding of the "One God" is quite similar in any religion that is in the general category of monotheism. This is because they all stem from the same Jewish sources. It is quite safe and accurate to say that regarding the concept of God himself we will find that Jews, Moslems, and Christians are pretty much in agreement. God is the all-Powerful, all-Knowing, Eternal, Transcendent Creator of the Universe, the Source of all being, who has created the universe and all that it contains, and "desires" to have a relationship with his creatures.

What emerges from this is that the existence of God and the details of a particular religious faith are two separate issues. It is quite possible to have a clear knowledge or belief in the existence of God the Creator while still having doubts about, or even rejecting, all Holy Scripture.

This book is not about Scripture or a particular religious dogma.

It is about the existence of God, the Creator. It is also about the proposition that God has created human beings with certain innate "hard wiring" that makes it clear to us that the purpose of our existence revolves around a relationship with this God. It may very well be true that without Divine revelation we are pretty much in the dark about what He may want from us or how we should relate to Him. That however does not change the fact that He created us and we are seeking Him. Within the parameters of this book I am a "missionary" for belief in God, not a particular religion.

Ground Rule #5: This Book Is Not about "Defending the Faith" of Judaism

Harris, Hitchens, Dawkins, and other atheist polemicists lambaste many different religions and their scriptures in their writings and speeches. As I made quite clear, this book is not about Scripture or a particular religion. Therefore, excepting the remarks I will make here and perhaps some tangential comments scattered throughout, I do not intend to put forth a comprehensive defense of Judaism against these vituperative railings. Suffice it to say the following:

Generally, these atheist writers display spectacular ignorance about what Judaism is, and most crucial of all, how Jews have understood and related to the Torah (Bible) for the past several thousand years. A pilot's manual for an F-15 fighter-bomber is not written with the purpose of enabling the average person to pick it up, read it, and attempt to fly an F-15. The manual is written with certain assumptions about the knowledge, background, and training of the person who is reading it. Without this knowledge, background, and training, there is no possible way for the uninitiated reader to have a proper understanding of what the manual is and what it is supposed to accomplish.

The Jewish Torah is no different. Speaking as an ordained Orthodox rabbi who has been a Jewish educator for three decades, I can categorically state the following: It is not possible to have a coherent or accurate understanding of neither the mission and nature of the Jewish people, Jewish law and

Jewish religious obligations, nor the *weltanschauung* of the Jew, by reading an English adaptation, of an English translation, of a Latin translation, of a Greek translation of a Hebrew Bible. The Torah as a **Jewish** Scripture does not mean whatever any particular reader thinks it means any more than the manual for the F-15 means whatever any particular reader thinks it means. Judaism has *its own* guidelines, parameters, and traditions regarding the study, interpretation, and applications of the Torah. Any explication of the Torah that is made in ignorance of or outside of these guidelines (no matter how fascinating, creative, or novel it may be) has nothing to do with Judaism. Judaism and its Torah are not beholden to or defined by the personal musings or speculative theories of a group of new-age atheists. If an individual wants to criticize the Torah worldview, he should first and foremost take the trouble to find out what it actually is. As I pointed out above, a superficial reading of an English translation of the text is, at best, a scratching of the surface of a real understanding of Judaism. To put it colloquially: the devil *can and does* quote Scripture for his own purpose.

As long as we are on the subject of the devil quoting Scripture, if awards were given out for grotesque distortions of Judaism by atheistic authors, Christopher Hitchens would be in a class by himself (ironically enough, he was Jewish). His obscene depiction of a Jewish circumcision ceremony, found in *God Is Not Great*,[4] could have been plagiarized from the pages of *Der Sturmer*, the rabidly anti-Jewish, quasi-pornographic tabloid published by Julius Streicher, a Nazi war criminal who was executed by the Allies in 1946. The way in which Hitchens portrays the ceremony is so vulgar and divorced from reality that one could justifiably wonder whether it might not have been a projection of some of his own darker unresolved issues. At the very least, it indicated some sort of pathology, be it emotional, spiritual, or intellectual. If that weren't sick enough, he also informed us (obviously after months of painstaking research on the subject) that Orthodox Jewish couples have intercourse through a "hole in the sheet."[5] It would have been closer to reality if he had informed us that the earth was flat. After all, at least from our limited perspective, the earth *looks* flat. What was his source for this information? It is obvious that he simply made it up. In light of the above, we can safely assume that unless presented with conclusive evidence to the contrary, anything negative that Christopher Hitchens wrote about Judaism is either a distortion, an outright fabrication, or presented so out of context that it is the equivalent of an outright fabrication.

In any case, if writers like Richard Dawkins or Victor Stenger want to read the Torah and understand it in their own distorted way, they should at least display some basic integrity and call it Stengerism or Dawkinsism, or any other name that suits them. Just don't call it Judaism.

Ground Rule #6: Regarding the "Leap of Faith"

It must be acknowledged that there is an important (valid) point that is raised by non-theists in their assault on religion. However, before I elaborate on that point, please note that I specifically wrote assault on *religion*, not assault on belief in God.

If someone has chosen to believe in a specific religious or non-religious system of values or any other items that fall in the category of "dogma," then it is perfectly valid to ask: How do you know it is true? You claim that your friend Sidney is the messiah? You claim he is a channel to the upper worlds? Prove it. You claim that embracing Socialism will raise the quality of life in a society? Prove it. To claim something is true because it can't be *disproven* is intellectually unacceptable. Does Dr. Sam Harris want me to accept his atheistic system of "morality" as valid? If so, he must *reasonably demonstrate* to me how his system could possibly be anything other than his totally subjective feelings on the subject. Similarly, if I want someone to accept the Ten Commandments as the word of God, that person is perfectly justified in asking for rational evidence to believe that they actually *are* the word of God.

There are those who say that an individual must take a "leap of faith" about certain fundamental core claims and *then* the truth of the proposition will become clear. If a fitting metaphor for discovering the truth about our existence is the slow, methodical, doggedly determined climb to the top of a tall mountain peak, then a leap of faith, in my opinion, is like jumping off the side of the mountain and expecting to miraculously end up at the summit. With a leap of faith, not only can you accept any type of ideological or religious dogma no matter how absurd, it is even possible, as we shall illustrate in the coming chapters, to believe that the first bacterium — which is more functionally complex and sophisticated than any machinery ever produced by human technology — could actually assemble itself without an intelligent designer. Put simply, with a leap of faith, it is possible to believe anything.

Ground Rule #7: Evolution – Irrelevant to the Discussion about the Existence of God

This book is *not* concerned with the following:

1. ***The Age of the Universe.*** There are a number of different authentic Jewish approaches to this subject, and a full presentation (besides being a distraction from our main thesis) is well beyond our scope here. Purely for the purposes of this book, and completely sidestepping the question of what Judaism's view is or isn't, we will work with mainstream cosmology and Big Bang theory that the universe came into existence roughly fourteen billion years ago.
2. ***The Truth or Non-truth of Darwinian Evolution and Natural Selection.*** We shall demonstrate in the coming chapters that contrary to popular belief, Darwinian Evolution is irrelevant to our question. None of this, of course, has anything at all to do with the topic of the Origin of Life. As Professor Richard Dawkins writes in *The God Delusion*: "Darwinian Evolution proceeds merrily once life has originated. But how does life get started?"[6] Yes, Professor Dawkins, how *does* life get started? This book is most definitely concerned with the Origin of Life.

End Notes

1 Heard by the author in a lecture given by Rabbi Noah Weinberg.

2 Ibid.

3 http://www.jewcy.com/dialogue/monday_why_are_atheists_so_angry_sam_harris

4 Christopher Hitchens, *God Is Not Great: How Religion Poisons Everything* (New York: Twelve, Hachette Book Group, 2009), p. 49.

5 Ibid., p. 54.

6 Dawkins, *The God Delusion*, p. 164.

Section 2

The Origin of Life: Clear Scientific Evidence of a Creator

Chapter 3

Yes, Professor Dawkins, How Does Life Get Started?

The conflict between skeptic and believer regarding the existence of God has been going on for much longer than many people realize. In Talmudic literature there are records of encounters between rabbinic sages and Greek and Roman skeptics that took place well over 2,000 years ago. One of the central questions in this ongoing debate is: How did life come to be on the Earth? The Book of Genesis clearly states that "in the beginning" life on Earth was the result of a Divine act of creation. The atheist counters that there is no need to invoke mythological fairy tales about a supernatural Creator. Rather, he argues, the enormous complexity and variety of life on our planet can be explained through scientific/naturalistic means.

In modern times, when believers and skeptics argue about science and religion, invariably the first subject that comes up is Darwinian Evolution and natural selection. I remain thoroughly unconvinced that evolutionary theory is anywhere near adequate to explain the organized complexity of the living world.* As stated earlier, however, it is clear to me that debates about the theory of evolution are beside the point, and for the purposes of this book the subject is a non-issue. In fact, for argument's sake, I am fully prepared to accept the fact of evolution of the species as explained by the Neo-Darwinian theory of one's choice. For our purposes, we would do well to simply bypass evolution. The critical issue that needs to be addressed is: *How did life begin?* Interestingly enough, after coming to this conclusion myself, I found that there are those on the other side who agree with me:

> If I were a creationist, I would cease attacking the theory of evolution...and focus instead on the origin of life. **This is by far the weakest strut of the chassis of modern biology**. The origin of life is a science writer's dream. It abounds with exotic scientists and exotic theories, which are never entirely abandoned or accepted, but merely go in and out of fashion.[1] (John Horgan, Senior Writer, *Scientific American*)

When Mr. Horgan writes that Origin of Life is "by far the weakest strut of the chassis of modern biology," what he means, of course, is that it is the weakest strut of an atheistic/materialistic approach to biology. That is to say, it is the weakest link in an attempt to explain the living world in purely naturalistic/materialistic terms, leaving God completely out of the picture. In truth, calling this strut "weak" is giving it way too much credit. As we shall see, it is a strut that *does not exist at all*, leaving the foundations of atheistic biology to be nothing but thin air and wishful thinking.

"In the Beginning" — Origin of Life and Darwinian Evolution Are Two Completely Separate and Unrelated Issues

Many people are under the impression that Darwinian Evolution is a neat scientific package that explains *everything* about life on this planet, including the beginning of life from the primeval "organic soup," the de-

---

* For example, see Dr. Stephen Meyer, *Darwin's Doubt (HarperCollins), the EvolutionNews.org website, and the "Scientific Dissent from Darwinism" petition signed by 800 scientists from all over the world, http://www.discovery.org/scripts/viewDB/filesDB-download.php?command=download&id=660.*

velopment of apes and human beings, and even why the Chicago Cubs have not won a World Series in over 100 years. This simply is not true. Origin of Life and Evolution of the Species are totally separate issues and must be investigated and understood in different ways (*nobody* will be able to explain the Cubs, though). It is important to note that this assertion is not a matter of scientific controversy.

Dr. Eugenie Scott, executive director of the fiercely pro-evolution National Center for Science Education (NCSE), writes in her book *Evolution vs. Creationism: An Introduction*:

> Although some people confuse the origin of life with evolution the two are conceptually separate. Biological evolution is defined as the descent of living things from ancestors from which they differ. Life had to precede evolution...We know much more about evolution than the origin of life.[2]

In other words, biological evolution can occur *only after life begins*. If "life precedes evolution" then evolution cannot explain the origin of life. The following metaphor illustrates the conceptual difference between Evolution and Origin of Life. During World War II, automobile assembly lines in the United States were reconfigured and retooled to produce half-tracks and tanks for the war effort. Roughly speaking, evolution would be comparable to the process of transforming an assembly line designed to produce cars into one that produces tanks. The basic machinery is already functioning and in place but needs modification, sometimes more and sometimes less, depending on what needs to be manufactured. Origin of Life, on the other hand, is analogous to the process of bridging the gap from raw materials (e.g., iron ore, copper ore, stone and gravel, sand, clay, trees, etc.) to actual building materials and machinery (e.g., bricks, cement, re-bar, steel beams, lumber, pipes, chains, glass, motorized pulleys, gears, etc.) and then assembling the building materials into a fully functioning manufacturing plant with an assembly line that produces automobiles.

It is clear that these are two fundamentally different types of tasks. At a later point we will re-explore these differences, discuss them in more precise detail, and explicate the far-reaching implications. For the time being, however, it is enough to understand that they are conceptually separate and require two separate scientific disciplines for their investigation. The study of Darwinian Evolution falls in the category of *Biology* — the study of living things, while the essential scientific discipline required for Origin of Life research is *Chemistry*. In other words, how did the original non-living, non-organic chemicals coalesce and form the first living cell, along with its complex molecular machinery, genetic information processing system, and the voluminous amount of information needed for life-sustaining processes and self-replication?

Origin of Life, God, and the Argument from Design

If our subject then is to be the Origin of Life and the existence of God from the viewpoint of the theist, the discussion will revolve around what has classically been called the "Argument from Design" in some form or another. This argument can be presented from different angles, and Richard Dawkins has even chosen to label it with a different name: the "Argument from Improbability." Be that as it may, both are essentially identical.

> ***Important note:*** *Of course I am aware that I am not the first to present a polemic on this topic. Still, I respectfully ask the reader to bear with me and follow carefully as I develop the argument and to please leave aside any pre-conceived notions one may have about this subject.*

The Argument from Design, though easy to understand, has been distorted and misrepresented in ways that have caused considerable confusion. I will alert the reader to these points of confusion as we go along. The argument, as I will present it, is extraordinarily simple. In fact, it is so simple that it is not really an argument at all; it is a given, a fundamental assumption about the way all human beings perceive reality. From this point on, anytime I mention the Argument from Design, I am referring to the following:

- The existence of my suit *itself* is the evidence and proof of the existence of the tailor who created the suit. I do not need to know what he looks like or where he lives, or any other independent evidence. The suit *itself* is all the evidence I need to be certain of the existence of the tailor.
- The existence of a poem *itself* on a piece of paper — for example, *The Charge of the Light Brigade* — is the indisputable evidence of the existence of the poet who authored the poem. The existence of my watch *itself* is the proof of the existence of the watchmaker or factory that created the watch.
- Suits, poems, and watches do not make themselves.
- The same applies to anything that we decide is in the category of a suit, poem, or watch.

It is critical for us to understand why we accept these propositions. It is critical for us to understand exactly *why* we accept as an incontrovertible truth that suits, poems, and watches do not make themselves.

First Point of Confusion: It Is Not Because of "Experience"

There is a common fallacy that the reason we assume that watches are the result of conscious, intelligent causation is because of *experience*. In other words, we know from *experience* that there are watchmakers and factories that produce watches, and therefore we conclude that *all* watches are produced by watchmakers and factories. This is a mistaken conclusion based on flawed logic. Our mere *experience* alone that *some/many* watches are created by watchmakers cannot in any way logically lead to the conclusion that *all* watches are created by watchmakers. The same applies to suits, poems, and for that matter, automobiles. (One might say that *one does not know* of any other process that assembles watches, but that would be an argument/conclusion based on ignorance.)

Philosopher David Hume (1711–1776), among others, has famously pointed out that our *experience* that the sun has risen in the east every day for thousands of years does not in any way constitute evidence that it will rise in the east tomorrow, or rise at all for that matter. That is to say, there is no logical/causative connection between the *mere experience* of the sun rising yesterday and the certainty of its rising tomorrow. (Therefore if we *are* certain that it will rise tomorrow — to the degree that human beings can be certain about anything — it must be for reasons other than just *experience.*)

In the same way, there is no logical/causative connection between our *experience* of *some* watches being manufactured and assembled by watchmakers and the proposition that *all* watches are manufactured and assembled by watchmakers. Based on our very limited *experience* alone, how could we possibly know how *all* the untold billions of watches and clocks in existence are assembled? For all we know, *some* watches might self-assemble through some as-yet-unknown process. It is a very tricky proposition indeed to draw absolute conclusions about our physical reality from mere *experience.*

Why then do we accept the Argument from Design (as stated above) as an undeniable truth? Why is our level of certainty that *all* watches are produced by watchmakers so high that it is at the same level of certainty as any mathematical proof? We are as certain that every watch in existence has been assembled by a watchmaker as we are certain that 2+2=4. In fact, we are so certain that all watches are produced by watchmakers that if someone seriously questioned that proposition we would seriously question their sanity.

The answer is really quite simple. We have not reached our conclusions based on *experience* alone. In light of the sum total of human experience and *observation*, in light of the totality of human experience and *repeated observation* that has taken place over thousands of years, we have extracted and formulated an *operative principle* that has been tested, applied, and found to be true without exception. It has been applied, observed, and tested so many times that it has the same status as any of the accepted laws of physics or chemistry: *There are certain levels of functional complexity, design, and specified information beyond which the human mind simply refuses to accept could be the product of an undirected process. Above these levels we take for granted the existence of an intelligent creative force that is the cause of these phenomena.* Watches, suits, and poems are all well beyond the levels of functional complexity and specified information required for us to conclude that they are the result of intelligent causation, and therefore we conclude with absolute certainty, in keeping with our principle, that *all* watches, suits, and poems are the result of intelligence.

Dr. Leonard Susskind, professor of theoretical physics at Stanford University, in his book *The Black Hole War*, outlines the operative principles of physics and mathematical probability involved (ironically, Dr. Susskind is an opponent of Intelligent Design theory):

> It would be a bad idea to park your BMW in the rainforest for five hundred years. When you came back, you'd find a pile of rust. If you left the pile of rust for another five hundred years, you could be pretty sure it wouldn't turn back into a working BMW. That in short is the Second Law of Thermodynamics... Both [the BMW and the rust heap] are collections of about 10^{28} atoms [10 multiplied by itself 28 times; in other words, the number 1 followed by 28 zeroes], mostly iron.
>
> Imagine that you took those atoms and threw them together randomly. What is the likelihood that they would come together to form a working automobile [i.e., a functionally complex machine]... I think we can all agree that it would be extremely unlikely... If you imagined all the possible ways that you could assemble the atoms, the overwhelming majority of the arrangements would look like rust heaps... If you shook up the atoms of a car, you would be much more likely to get a pile of rust because there are so many more rust pile arrangements than car arrangements. [3]

If we subjected the atoms or parts of a watch or suit to undirected, random forces, we would almost certainly end up with a meaningless arrangement of junk. This is because there are so many more junk arrangements than suit or watch arrangements. That is why we are absolutely certain that *all* watches are assembled, not by undirected, random forces, but by an intelligent creative force called a watchmaker, *all* suits by a tailor, and *all* BMWs by an automobile assembly line. Dr. Susskind explains how the same principle/law (not mere *experience*) applies to specified information, such as the poem we mentioned above:

> Here is another example. An ape banging away at a typewriter will almost always type gibberish. Very rarely will he type a grammatically correct sentence...What's more, if you take the letters of a meaningful sentence and shake them up like tiles in a Scrabble game, the result will almost always be gibberish. The reason? There are a lot more nonsensical ways to arrange twenty or thirty letters than there are meaningful ones.[4]

Dr. Robert Shapiro (1935–2011), professor emeritus of chemistry at NYU (a self-declared agnostic and opponent of Intelligent Design theory), calculated that the odds of a chimp banging away at a computer keyboard typing out the sentence "To be or not to be that is the question" (including spaces), is one chance in 10^{66} (10 multiplied by itself 66 times); this number is so large that it approaches the number of atoms estimated to be in the Milky Way galaxy.[5] To put this into perspective, the number of seconds that have elapsed since the Big Bang some fourteen billion years ago is approximately 4×10^{17}. This is a laughably small number when compared with 10^{66}. Even if we would grant the impossible scenario of the chimp typing out a new combination of forty letters every second from the moment of the Big Bang until now, his chances of emulating Shakespeare would still be virtually zero (a little more than one chance in 10^{49}). In other words, the occurrence of such an event is as close to impossible as we can imagine.

We now understand why Dawkins chose the term "Argument from Improbability." It is therefore clear that when we assert that functionally complex objects like suits and watches and examples of specified, coherent information like poems do not make themselves and can only be the results of intelligent intervention, we are on very safe ground.

The levels of functional complexity, design, and specified information that we observe in the living world are quantum leaps beyond anything we've mentioned so far. The organ we call a "heart" is nothing more and nothing less than an astoundingly efficient electric pump. (If we would subject the atoms or parts of a heart to random, undirected forces, we would end up with a pile of organic goo.) We could mention the sonar of bats, the navigational system of birds and fish, the electrical generating apparatus of an eel — the list is endless.

It would seem reasonable to conclude that living organisms are also the result of the intervention of an intelligent creative force. In fact, for a good portion of human history, this conclusion was considered to be eminently reasonable. The theist still considers it to be true while the atheist rejects it. We need to discover *precisely* the point of contention that brought about this intellectual/ideological divorce. In order to achieve this it is necessary to focus on two more points of confusion surrounding the Argument from Design.

Two Critical Points of Confusion

- Point number 1: Contrary to popular belief, *nobody* disagrees about the existence of very high levels of design, functional complexity, and specified information in the living world. I will present ample evidence below to support this assertion.
- Point number 2: Contrary to popular belief, *nobody* disagrees with the Argument from Design, including every atheist and skeptic in the world. That is to say, there is no one in his right mind who does not agree that the suit proves the existence of the tailor, the poem the existence of the poet, and the watch the existence of the watchmaker. Likewise, there is no one in his right mind who does not agree that suits, poems, and watches do not make themselves. There is no one who disagrees with the fundamental laws of physics and mathematical probability that are behind the truth of these propositions. Where then do we disagree?

Where skeptic and believer part ways is not regarding the existence of design in the living world, nor regarding the validity of, and the aforementioned philosophical/scientific underpinnings that validate, the Argument from Design. The point of divergence is that the atheist/non-believer argues that the *Argument from Design cannot be applied to the world of living systems, that it does not apply to the design observed in living organisms.* To the skeptic, the Argument from Design does indeed prove a creator of the sonar of a nuclear submarine, which uses it to hunt enemy subs thousands of feet beneath the ocean's surface. However, to that same skeptic, it does not prove a creator of the sonar of the sperm whale, which uses it to hunt giant squid thousands of feet beneath the ocean's surface. Why not?

The atheist claims that for a very unique and specific reason living organisms are an exception to the Argument from Design; that incredibly, enough scientists have discovered a naturalistic, unguided process that *mimics design*; a process that creates the powerful *illusion* of design and designer! What is this process that allegedly creates functional complexity and generates specified information without the aid of a conscious, intelligent, and creative force? The process called Darwinian Evolution:

- **Professor Richard Dawkins, Oxford University**: (A) "Charles Darwin discovered a way in which the unaided laws of physics could, in the fullness of geological time, come to mimic design."[6] (B) "Biology is the study of complicated things that give the appearance of having been designed for a purpose."[7]
- **Dr. Jerry Coyne, Professor of Evolutionary Biology, University of Chicago**: "If anything is true about nature, it is that plants and animals seem intricately and almost perfectly designed for living their lives. Squids and flatfish change color and pattern to blend in with their surroundings...bats have radar to home in on insects at night... hummingbirds are far more agile than any human helicopter...what does all this seem to imply? A master mechanic, of course."[8]
- **Dr. Lawrence Krauss, Physicist, Arizona State University**: "The illusion of purpose and design is perhaps the most pervasive illusion about nature that science has to confront on a daily basis. Everywhere we look, it appears that the world was designed so that we could flourish."[9]
- **Dr. Francisco Ayala, Professor of Evolutionary Biology and Philosophy, University of California, Irvine**: Title of an article he authored in May/2007 for the National Academy of Sciences: "Darwin's Greatest Discovery: Design without Designer."
- **Dr. Francis Crick, Physicist and Molecular Biologist, awarded the Nobel Prize for Medicine in 1953 (together with Dr. James Watson) for discovering the molecular structure of DNA**: "Biologists

must constantly keep in mind that what they see is not designed but rather evolved."[10]

- **The Wyss Institute at Harvard University**: "The Wyss Institute for Biologically Inspired Engineering uses Nature's *design principles* to develop bio-inspired materials and devices that will transform medicine and create a more sustainable world... The Wyss Institute aims to discover the *engineering principles* that Nature uses to *build living things*."[11]

In other words, this blind process called Darwinian Evolution creates what looks like incredibly sophisticated, designed systems in the living world, hence the name of Richard Dawkins' bestselling book about evolution, *The Blind Watchmaker*. My response as a theist? As I noted earlier, I see no point here in getting involved in the battle over evolution and therefore I am prepared to simply concede its truth. Solely for the purposes of this book — i.e., for argument's sake — we will accept Darwinian Evolution as scientific fact.*

One might mistakenly conclude that accepting the fact of evolution marks the end of our discussion about God the Creator. In fact, it is only the *beginning* of the discussion. What I will demonstrate in the following pages is that Darwinian Evolution, which is the foundation-stone upon which rests any "science-based" atheistic worldview, is totally and completely irrelevant to the question of the existence of a Creator of life. Even if we grant Darwinian Evolution the highest preferred status in the world of scientific theory, it provides (at best) nothing more than an *illusion* of scientific credibility for atheism.

Darwinian Evolution is based on random mutations in the replication of the DNA of a living organism. Sometimes by pure luck these mutations confer a survival advantage to the particular organism, which in turn makes it more likely that these traits will be passed on to future generations. Eventually, over millions, tens of millions, and hundreds of millions of years, all these little changes add up and from the original mi-

---

* To avoid confusion or appearances of disingenuousness, I reiterate that in actuality I do not accept Darwinian Evolution as an adequate explanation for the stunning variety and complexity that exists in the living world.

croscopic bacteria, here we are! However, even going on our assumption-for-argument's-sake that the theory is true, it is all based on an existing, self-replicating living organism with a fully functioning DNA-based genetic code. Darwinian Evolution and Natural Selection are only operative and relevant from that point forward. Neo-Darwinian theory does not even pretend to account for the existence of the first self-replicating bacterium. As we have pointed out, how that living cell got there in the first place is the subject of the fundamentally and conceptually separate scientific area of research called Origin of Life.

In truth, no scientist alive today would claim that evolution started with bacteria, the "simplest" and oldest known life form. As we shall see, there is nothing "simple" about a bacterium. All agree that the notion of a fully functioning DNA-based bacterium suddenly popping out of a pre-biotic swamp on the ancient earth through *natural means* is preposterous. Why? For the exact same reason that the notion of a watch, poem, suit, or mechanically sound BMW suddenly popping out of a swamp in the rainforest is preposterous. A bacterium and its genetic machinery are simply too functionally complex and sophisticated to emerge fully assembled due to some lucky accident.

Dr. Robert Shapiro (among others), in his classic work, *Origins: a Skeptics Guide to the Creation of Life on Earth*, discussed the odds for the spontaneous self-assembly of the simplest living organism known to have ever existed: a bacterium. In order to illustrate the patent absurdity of such a notion, he sets up an impossibly favorable scenario in order to maximize the possibility of the random occurrence of such an event. The hypothetical set up is as follows:

- The time allowed for each attempt at spontaneously assembling a bacterium from simple chemicals is one minute.
- The total time allotted for these attempts is one billion years, much longer than the actual maximum amount of time available for the first living cells to form on earth.
- The available space is *the entire earth covered by an ocean ten kilometers deep!*
- That entire space is divided into tiny reaction chambers of bacterial

size: one square micrometer apiece (micrometer = one millionth of a meter).

- Every minute for one billion years, a new attempt is made to spontaneously assemble a bacterium in each of these chambers.

Even given such an unrealistically favorable set of circumstances Dr. Shapiro comes to the following conclusion:

> Harold Morowitz, a Yale University physicist...has calculated the odds [against success in a single trial]...as one chance in 10 to the 100 billionth power... The improbability involved in generating even one bacterium is so large that it reduces all considerations of time and space to nothingness...we would truly be waiting for a miracle.[12]

Many scientists, therefore, offer as a *point of conjecture and speculation* that a purely naturalistic origin of life must begin with some form of simple self-replicating molecule that would somehow evolve into a fully functioning living cell. No one has ever seen such a molecule and whether or not it has ever existed is itself an astoundingly improbable speculative proposition. Origin of Life researchers have recently *designed*, *constructed*, and *synthesized* RNA* molecules that in *an extremely limited* use of the term could be described as self-replicating. This breakthrough was achieved by some of the world's leading researchers using state-of-the-art equipment, following scrupulously controlled manufacturing processes and protocols, and under the most rigorous of laboratory conditions; in other words, conditions that bear a striking resemblance to an automobile assembly line (intelligent design), but no resemblance at all to nature.

To conclude that because intelligent designers (i.e., expert scientists) can manufacture something resembling a self-replicating molecule in the laboratory, it is therefore plausible that such a molecule could result from natural processes, is as absurd as proposing that because a GM plant in

---

* DNA contains the actual genetic information for the organism. RNA is a "simpler" macromolecule that plays a crucial role in the replication of DNA and the synthesis of proteins by the living cell.

Michigan can produce a Chevy Impala, it is therefore plausible for a Chevy Impala to emerge through natural processes. As one world-class chemist put it: "Unfortunately neither chemists nor laboratories were present on the early earth to produce RNA."[13] (For the sake of continuity and clarity, the issue of the "simple" self-replicating molecule is addressed separately in Appendix 1.*)

Before we draw our final conclusions about evolution and its place in the quest to discover the truth about the existence of God, it is critical to understand how formidable a problem we are facing when we speak about the formation of the first living bacterium and its genetic code.

A Leading Atheistic Scientist Describes the Genetic Information Processing System

Professor Richard Dawkins, born in Kenya (then a British colony) in 1946, is the "grand old man" of contemporary atheism. He is held in papal-like awe by ideologically committed atheists all over the world. Educated at Oxford University, he has held prestigious academic positions in both the United States and Great Britain. Two of his most well-known books are *The Blind Watchmaker: Why Evidence of Evolution Reveals a Universe without Design* (1986), and *The God Delusion* (2006). Here is how Dawkins describes the intricate functioning of genetic coding in the "simplest" of living cells:

> After Watson and Crick we know that genes themselves...are long strings of pure digital information. What is more they are truly digital, in the full and strong sense of computers and compact discs. The machine code of the genes is uncannily computer-like. Apart from differences in jargon, the pages of a molecular biology journal might be interchanged with those of a computer engineering journal. Our genetic system, which is the universal system for all life on the planet is digital to the core...DNA characters are copied with an accuracy that rivals anything modern engineers can do...DNA messages are...pure digital code.[14]

---

* In Appendix 1, we also discuss the famous "spark discharge" experiment conducted by Dr. Stanley Miller in 1953.

> I'm fascinated by the way molecular genetics has become a branch of information technology...genetics [is] digital, high-fidelity, a kind of computer science.[15]

The late astronomer and science writer, Dr. Carl Sagan:

> A living cell is a marvel of detailed and complex architecture... The information content of a simple cell has been estimated as around 10^{12} bits, comparable to about a hundred million pages of the Encyclopedia Britannica.[16]

J. Craig Venter, one of the world's leading synthetic biologists informs us that:

> All living cells that we know of on this planet are DNA-software-driven biological machines comprised of hundreds of thousands of protein robots — coded for by [digital information stored in the] DNA — that carry out precise functions. We are now using computer software to design new DNA software.[17]

Dr. George Church, a molecular geneticist at the Wyss Institute at Harvard University, has done pioneering work on the development of new technologies to synthesize DNA, along with utilizing DNA as a feasible storage medium for our own needs. Regarding the digital information storage capacity of DNA, he writes that "DNA is among the most dense and stable information media known."[18] How dense? One gram of DNA — the weight of two Tylenol tablets — can theoretically store 455 exabytes of information.[19] In layman's terms: one gram of DNA can store the same amount of digital information as *one hundred billion* DVDs. One large test tube or beaker of DNA could store all the information that exists on the Earth today, from every library, database, and computer in the world. In short, the DNA information processing system that operates in every living thing in existence is the most sophisticated and advanced digital information system known to man. Now we understand what prompted Bill Gates to write in his book, *The Road Ahead*: "DNA is like a computer program, but far, far more advanced than any software we've ever created."[20]

Origin of Life expert, Dr. Paul Davies of Arizona State University, sums up the dilemma faced by researchers:

> In a living organism we see the power of software, or information processing, refined to an incredible degree...the problem of the origin of life reduces to one of understanding how encoded software emerged spontaneously from hardware. How did it happen? How did nature "go digital?"[21]

If DNA is a digital information storage system loaded with vast amounts of information, what exactly is this information used for? The short, simplified answer is as follows: The digital information stored in DNA is accessed, read, and copied by the cell's sophisticated molecular machinery, including different types of RNA. It is then translated by other types of RNA and molecular machinery in order to produce the aforementioned "protein robots" that perform the necessary functions for the cell's survival. Dr. James Shapiro, a University of Chicago microbiologist, has stated that the "biochemical, structural, and behavioral complexities" of this system that operate in the "simplest" bacteria are so incredible that they "outstrip scientific description."[22] Yes, how *did* nature go digital?

The Universe Contained within a Millionth of a Meter

It is important that we not be fooled into thinking that the size of a bacterium, which is measured in the millionths of a meter, is an indication of its simplicity. Dr. Jack Szostak, a Nobel Prize-winning Origin of Life researcher (and staunch opponent of Intelligent Design theory), has written the following:

> Every living cell, even the simplest bacterium, teems with molecular contraptions that would be the envy of any nanotechnologist. As they incessantly shake or spin or crawl around the cell, these machines, cut, paste and copy genetic molecules, shuttle nutrients around or turn them into energy, build and repair cellular membranes, relay mechanical, chemical or electrical messages — the list goes on and on...it is virtually impossible to imagine how a

> cell's machines, which are mostly protein-based catalysts called enzymes, could have formed spontaneously as life first arose from nonliving matter around 3.7 billion years ago.[23]

Dr. Bruce Alberts, biochemist and former president of the National Academy of Sciences:

> The entire cell can be viewed as a factory that contains an elaborate network of interlocking assembly lines, each of which is composed of a set of large protein machines... Why do we call the large protein assemblies that underlie cell function protein machines? Precisely because, like machines invented by humans to deal efficiently with the macroscopic world, these protein assemblies contain highly coordinated moving parts.[24]

As we stated earlier, there is nothing simple about the bacterial cell; on the contrary, the *miniaturization* of the "molecular contraptions," the "DNA software driven biological machines," and the thousands of "protein robots," are an indication of its *sophistication* rather than simplicity. Dr. Richard Strohmam, professor emeritus of cell biology at UC Berkeley puts it this way:

> Molecular biologists and cell biologists are revealing to us a complexity of life that we never dreamt was there. We're seeing connections and interconnections and complexity that are mind-boggling. It's stupendous. It's transcalculational[*sic*]. It means that the whole science is going to have to change.[25]

Let's briefly explore the amazing world of the bacterium. The following passages are from the works of the aforementioned Dr. Paul Davies and Dr. Michael Denton, an Australian geneticist and microbiologist:

> The living cell is the most complex system its size known to mankind. Its host of specialized molecules...execute a dance of exquisite fidelity, orchestrated with breathtaking precision. How can mindless molecules, capable of only pushing and pulling their immediate neighbors, cooperate to form something as ingenious as a living organism? [26]

> Scientists have fabricated invisible cogwheels, motors the size of a pinhead, and electrical switches as tiny as individual molecules... the burgeoning field of nanotechnology — building structures and devices measured on a scale of billionths of a meter promises to revolutionize our lives...but we should not lose sight of the fact that nature got their first. The world is already full of nanomachines: they are called living cells. Each cell is packed with tiny structures that might have come straight from an engineer's manual. Miniscule tweezers, scissors, pumps, motors, levers, valves, pipes, chains, and even vehicles abound.[27]
>
> Although the tiniest living things known to science, bacterial cells, are incredibly small, each is a veritable micro-miniaturized factory containing thousands of elegantly designed pieces of machinery... more complicated than any machine built by man and without parallel in the non-living world. What we would be witnessing would be an object resembling an immense automated factory. However it would be a factory which would have one capacity not equaled in any of our own most advanced machines, for it would be capable of replicating its entire structure.[28]

Dr. Denton goes on to describe the cell as "the complexity of a jumbo jet packed into a speck of dust invisible to the naked eye... Moreover it is a speck-sized jumbo jet which can duplicate itself quite easily."[29] Dr. Graham Cairns Smith of the University of Glasgow also expresses his wonderment at the self-replicating ability of the smallest living:

> It may seem hardly surprising that no one has ever actually made a self-reproducing machine, even though Von Neumann laid down the design principles more than forty years ago. You can imagine a clanking robot moving around a stock room of raw components choosing the pieces to make another robot like itself. You can show that there is nothing logically impossible about such an idea: that tomorrow morning there could be two clanking robots in the stock room (I leave it as a reader's home project to make the detailed engineering drawings).
>
> There is nothing clanking about E. coli [a type of bacteria found

> in the human stomach]; yet it is such a robot and it can operate in a stock room that is furnished with only the simplest raw components. Is it any wonder that E. coli's [DNA molecular] message tape is so long? (If you remember, the paper equivalent would be about ten kilometers long.) Is it any wonder that no free-living organisms have been discovered with [DNA molecular] message tapes below two kilometers? Is it any wonder that Von Neumann himself, and many others, have found the origin of life to be utterly perplexing?[30]

A Xerox machine can make a perfect copy of a color photograph. What we are talking about here is a Xerox machine that *makes another Xerox machine*. What about an automobile that makes another automobile or an F-15 fighter-bomber that makes another F-15 fighter-bomber? These are feats that are beyond our wildest engineering dreams and yet it is happening all around us trillions and trillions of times a day on a microscopic level. There we have it, a description of the "simplest" living cell and its genetic code in a nutshell...a millionth of a meter nutshell.

The Final Point of Confusion: The Myth that "Simple" or "Primitive" Life Has Ever Existed

"The essential problem in explaining how life arose is that even the simplest living things are stupendously complex."[31] — Dr. Paul Davies

It is common for people to talk about the development of complex life forms from *simple* living organisms or higher animals like mammals and primates evolving from *primitive* forms of life. The use of these types of terms is an egregious and inexcusable distortion of the reality. It only adds insult to injury that it is mainly scientists — who supposedly pride themselves on their pinpoint precision terminology — who are the truly guilty parties in the perpetuation of this distorted view. With this understanding then we clear up the final point of confusion about the Argument from Design: *there is no such thing as simple or primitive life.* No human being has ever seen such a thing, nor is there evidence that "simple" or "primitive" life has ever existed:

"We must concede that bacteria — prodigiously complex biochemical factories, endowed with a system of self-replication, containing DNA and RNA, themselves very complex molecules — **have no identified ancestors**."[32] — Dr. Alexandre Meinesz, professor of biology, University of Nice – Sophia Antipolis

In the same vein, Richard Dickerson, a molecular biologist at UCLA, wrote the following in *Scientific American*: "The complex genetic apparatus in present-day organisms is so universal that one has few clues as to what the apparatus may have looked like in its most primitive form."[33] Note that Dr. Dickerson has made the purely speculative assumption that at some unknown time in the past there was a "primitive" apparatus in existence. Why make an assumption that completely lacks any supportive evidence? Perhaps the obvious reason why we have so few clues as to the "primitive form" of the complex genetic apparatus that operates in every living organism is because there never was a primitive form. By the same token, the simplest explanation why bacteria have no "identified ancestors" is because in fact they have no ancestors.

The *Encyclopedia of Evolution* puzzlingly notes: "The oldest known remains of life from the fossil record appear surprisingly complex."[34] Only "surprising" if the unjustified assumption is made that there was anything before. "Surprisingly complex" is in fact a gross understatement. What has been discovered to have existed in the past and what exists today is *only* awesomely and staggeringly complex and sophisticated life — an accurate description of a bacterium — and other forms of life that are even *more* awesomely and staggeringly complex and sophisticated.

Dr. Stuart Kaufmann, a highly accomplished and respected scientist and Origin of Life researcher, cogently comments on this very point:

> **The [problem] I find most insurmountable is the one most rarely talked about**: all living things seem to have a minimal complexity below which it is impossible to go [i.e., below a certain level of complexity you don't get primitive life, you get death]... all free-living cells have at least the minimum molecular diversity of pleuromona [i.e., the "simplest" types of bacteria; for instance a *mycoplasma genitalium*]. Your antenna should quiver a bit here. Why is there this minimal complexity? Why can't a system simpler than a pleuromona be alive?[35]

What does he mean by the phrase: "your antenna should quiver a bit here"? In layman's terms: It should blow your mind that the "simplest" living cells are so mind-bogglingly complex and sophisticated. Why indeed can't a system simpler than a bacterium be alive? How in the world was the gap between non-life and life crossed through a purely unguided, naturalistic process? Does anyone other than myself find it interesting that the "most insurmountable" problem in discovering a naturalistic origin of life is the one "most rarely talked about"? If one's mission is to discover the truth, it is very strange indeed. However, if the point is not to rock the boat with embarrassing facts, it makes perfect sense.

What then can we conclude about the relationship between Darwinian Evolution and the question of the existence of a Creator? There *is* no relationship — it has no relevance at all. Darwinian Evolution simply begs the question with a vengeance. The very best that Darwinian Evolution can tell us is that once you have in place a fantastic piece of functionally complex molecular machinery called a living bacterium, which has at its core a highly sophisticated self-replicating system based on the storage, retrieval, and translation of encyclopedic amounts of digitally encoded information, then the interactions between this nanotool-filled organism, its "uncannily computer-like" genetic code, and its environment, are able over millions of years to morph into and produce an astounding variety of forms of organic machinery. All varieties of life are possible, if — and only if — this molecular machinery is in place. Where did it come from?

The Myth of "Undirected Natural Processes"

I have a close friend who has a beautiful built-in pool in his back yard. He owns an electrically powered device about the size of a laptop computer that meanders along the bottom of the pool, cleaning as it goes. It hits a wall and heads back in a slightly different direction, eventually cleaning the entire bottom of the pool. Imagine we are standing on his deck watching and he turns to me and says, "This is amazing, I'm standing up here doing absolutely nothing and the bottom of my pool is getting cleaned by an entirely undirected process!" I reply, "Undirected process?! That's not an undirected process...where did that machine come from?" (In fact, if Professor Richard Dawkins had written a book about it, he would have called it *The Blind Pool Cleaner.*) For those who claim that evolution is an "undirected process," I would offer a simple retort: That's no undirected process...where did that machinery come from?

Lest anyone have the impression that the profoundly significant nature of this line of reasoning can only be appreciated by those with inclinations toward religion, here is distinguished professor of philosophy at NYU, Thomas Nagel (a self-proclaimed atheist who describes himself as being "just as much an outsider to religion as Richard Dawkins"):

> The entire apparatus of evolutionary explanation therefore depends on the prior existence of genetic material with these remarkable properties...since the existence of this material or something like it is a precondition of the possibility of evolution, evolutionary theory cannot explain its existence. We are therefore faced with a problem...we have explained the complexity of organic life in terms of something that is itself just as functionally complex as what we originally set out to explain. So the problem is just pushed back a step: how did such a thing come into existence?[36]

Dr. Paul Davies states the same conundrum in the following manner:

> The replicative [sic] machinery of life is based on the DNA molecule, which is itself as structurally complicated and intricately arranged as an automobile assembly line. If replication requires such a high threshold of complexity in the first place how can any replicative system have arisen spontaneously?[37]

Dr. Addy Pross, professor of chemistry at Ben Gurion University in Israel:

> Despite the widespread view that Darwinian Evolution has been able to explain the emergence of biological complexity that is not the case... Darwinian theory does not deal with the question how [life] was able to come into being. The troublesome question still in search of an answer is: **How did a system capable of evolving come about in the first place**?[38]

As it turns out, even if we accept its truth for argument's sake, Darwinian Evolution is not, as the atheist would have us believe, a testimony to what can emerge from undirected natural processes. It is a testimony to the unimaginably awesome capabilities and potentials contained in the first living cell and its genetic code. From this, a paradigm shifting insight emerges: Contrary to popular belief, not only is Darwinian Evolution *not* the cause or explanation of the staggering complexity and sophistication of life on this planet, Darwinian Evolution itself is a process which is the *result* of the staggering complexity and sophistication of life on this planet. That is to say, once you have in place fantastic machinery, fantastic things can happen. However, the notion that fantastic machinery can do fantastic things was never in dispute. The only relevant question is where does fantastic machinery come from in the first place? How did the machinery of life come into being?

Dr. Robert Shapiro expressed the enigma in the following manner: "I am not an expert in evolutionary theory but have no reasons to quarrel with the conclusions of my scientific colleagues who are better informed. I feel however that the Origin of Life is a topic which is *much more fundamental* to the debate over intelligent design. The difference between a mixture of simple chemicals and a bacterium is much more profound than the gulf between a bacterium and an elephant."[39] Renowned biologist Dr. Lynn Margulis made an almost identical observation: "To go from bacterium to people is less of a step than to go from a mixture of amino acids to a bacterium."[40]

In other words, the development and evolution of different life-forms could be viewed as nothing more than grand variations on a theme. On the other hand, the jump from non-life to life cannot in any way at all be viewed as a variation on a theme. In the words of Dr. Michael Denton: "Between a living cell and the most highly ordered non-biological system such as a crystal or snowflake there is a chasm as vast and as absolute as it is possible to conceive."[41] As we have pointed out over and over again, non-life to life is a radically and fundamentally different type of transition — like going from lead to gold — and this is why it presents such a powerful challenge to the skeptic and non-believer.

Let us put it a different way: Darwinian Evolution in no way at all confronts the questions raised by, and the conclusions drawn from, the Argument from Design. The Argument from Design raises the following question: Where does functionally complex machinery and specified information come from? It answers the question by concluding that from human experience, repeated testing and observation, and in conjunction with fundamental principles of physics and mathematical probability, functionally complex machinery and specified information are *always* the result of intelligent causation. Darwinian Evolution is at best nothing more than a description of the workings of some of the most sophisticated machinery and information processing systems in existence, *without any meaningful attempt to explain their origin*. The irony that is lost on non-theists is that it is *only* the explanation of their origin that could address and refute the Argument from Design.

It may seem hard to believe (and indeed for many non-believers will be a bitter pill to swallow), but it becomes clear then that the prodigious efforts of evolutionary biologists/atheistic ideologues like Richard Dawkins, Jerry Coyne, and P. Z. Myers to deny the existence of a creator and intelligent design by triumphantly touting Darwinian Evolution to explain the functional complexity of living organisms have resulted in nothing more than a grand tautology: they have explained the amazing complexity of living organisms in terms of the amazing complexity of living organisms.

Evolution might possibly raise questions about the first chapter of Genesis, but that is not our concern here. The only question that matters for us is the one that Dawkins poses in *The God Delusion:* "Darwinian Evolution proceeds merrily once life has originated, but how does life get started?" Let's pose this question to those who are considered to be the elite of the atheistic/non-believing world: scientists. What will become eminently clear is that the scientific information necessary to counter the Argument from Design is the very scientific information that is glaringly absent.

Testable and Falsifiable Predictions of the Argument from Design

Before we examine what scientists can offer us in terms of understanding an unguided/naturalistic emergence of life, please consider the following thought problems. These will put the current scientific positions on Origin of Life in their proper perspective. It will also enable us to make testable/falsifiable predictions based on the Argument from Design.

- If someone told you he was undertaking an investigation into how/if it were plausible to go from the abundantly available raw materials of mud, stones, straw, and leaves, to a circle of mud huts in a clearing in a jungle through an entirely unguided, naturalistic process, what results would you predict? The answer is obvious: the proposition itself is absurd and we would correctly predict absolute failure.
- If someone told you he was undertaking an investigation into how/if it were plausible to go from the raw materials of thousands of sharpened pencils mixed with thousands of sheets of blank paper to the first stanza of the poem *The Charge of the Light Brigade* through an entirely unguided, naturalistic process, what results would you predict? The answer again is obvious: the proposition itself is so absurd that we would correctly predict failure.
- I once bought a boxed set of LEGO blocks for my children that could be assembled into a rather large pirate ship that appeared in full color on the front of the box. The "building blocks" for this pirate ship were abundantly available; in fact, the entire box was filled with hundreds of building blocks *designed specifically* to construct the ship. If someone proposed an investigation into how/if it were plausible for the ship to self-assemble through an unguided process, would anyone in their right mind predict anything other than utter failure for such a nonsensical idea?
- If someone then informed you they were undertaking an investigation into how/if it were plausible to go from a random mixture of non-living, non-organic chemicals to a nanotechnology-filled molecular machine that self-replicates by means of the most sophisticated digital information storage, retrieval, and translation system in existence on Earth —

encoded with encyclopedic amounts of specified information that direct the synthesis of all the molecular machinery needed to sustain the organism — through an entirely unguided naturalistic process, what results would you predict? *In fact, the results are exactly what we would expect and predict for an investigation into such an absurd and outrageous proposition.*

The "Dirty Secret": The Origin of Life Field Is a Failure

Consider the following words of Dr. Eugene Koonin, molecular biologist and highly respected Origin of Life researcher:

> The origin of life is one of the hardest problems in all of science... Origin of Life research has evolved into a lively, interdisciplinary field, but other scientists often view it with skepticism and even derision. This attitude is understandable and, in a sense, perhaps justified, given the "dirty" rarely mentioned secret: Despite many interesting results to its credit, when judged by the straightforward criterion of reaching (or even approaching) the ultimate goal, the Origin of Life field is a failure — we still do not have even a plausible coherent model, let alone a validated scenario, for the emergence of life on Earth. Certainly, this is due not to a lack of experimental and theoretical effort, but to the extraordinary intrinsic difficulty and complexity of the problem. A succession of exceedingly unlikely steps is essential for the Origin of Life...**these make the final outcome seem almost like a miracle.**[42]

It is worth noting that Dr. Koonin is not the only scientist who brings up the subject of "secular" miracles:

Dr. Francis Crick: "An honest man, armed with all the knowledge available to us now, could only state that in some sense, the origin of life appears at the moment to be **almost a miracle**."[43]

Dr. Graham Cairns-Smith: "Now I cannot deny all these possibilities: **that life on the Earth may be a miracle**, or a freak, or an alien infection. And I agree that the confidence was misplaced that supposed in the fifties that the answer to the origin of life would appear in some footnote to the answer to the question of how organisms work..."[44]

"It is not just the sheer size of even the smallest libraries [of the simplest organisms], it is not just that nucleotide units are rather complex... and difficult to join together...it is not just the need for enzymes...it is not just that ribosomes are so very sophisticated... There seems to be a more fundamental difficulty. Any conceivable kind of organism would have to contain messages of some sort and equipment for reading and reprinting the messages; any conceivable organism would thus have seem to have to be packed with machinery and as such **need a miracle (or something)** for the first of its kind to have appeared."[45]

Dr. Paul Davies: "...the various components fit together to form a smoothly functioning whole, like an elaborate factory production line — the **miracle** of life is not that it is made of nanotools, but that these tiny diverse parts are integrated in a highly organized way."[46]

"Abiogenesis [life from non-life] strikes many as virtually **miraculous**... It is even conceivable that scientists will one day create life of some sort in the laboratory, and thus...demonstrate convincingly that a **miracle** isn't needed."[47]

"You might get the impression from what I have written not only that the origin of life is virtually impossible, but that life itself is impossible... fortunately for us, our cells contain sophisticated chemical-repair-and-construction mechanisms, handy sources of chemical energy to drive processes uphill, and enzymes with special properties that can smoothly assemble complex molecules from fragments... But the primordial soup lacked these convenient cohorts of cooperating chemicals... So what is the answer? Is life a **miracle** after all?"[48]

Dr. Euan Nisbet, Professor of Geology, University of London: "Life is improbable, and it may be unique to this planet, but nevertheless it did begin, and it is thus our task to discover how the **miracle** happened."[49]

I'm quite aware that none of these scientists believe life came about by a Divine act of creation. However, when confronted with their own paucity of real knowledge of how life started and the lack of clear direction to scientifically find a solution, they find that the word "miracle" becomes a very handy, descriptively accurate, and a highly appropriate noun to use when describing the dizzying complexity of the simplest living organisms. That in itself is something that any truth-seeking individual needs to ponder carefully.

In February, 2011 there was a much publicized *Origins* conference at Arizona State University.[50] The panel at the conference contained an all-star line-up of scientists, including Professor Richard Dawkins, Dr. Paul Davies, J. Craig Venter, Dr. Christopher McKay of NASA, Dr. Lawrence Krauss, and Nobel Laureates Sidney Altman and Lee Hartwell. The results and outcomes stemming from the conference were *exactly* what we would have expected and predicted above.

John Horgan, reporting on the conference for *Scientific American*, summed it up by penning an article with the following title: "*Psst! Don't tell the creationists, but scientists don't have a clue how life began.*"[51] Mr. Horgan's tongue-in-cheek exhortation is of course superfluous; what follows below clearly illustrates that the secret's been out for quite a while. It also clearly illustrates that our predictions based on the Argument from Design have come true in spades. For some odd reason, though, the public at large still seems unaware of the true state of affairs in the Origin of Life field and even more unaware of the implications of that fact.

(1859) "On the Origin of Species by means of Natural Selection" by Charles Darwin is published: "Significantly, Darwin himself explicitly avoided the origin of life question, recognizing that within the existing state of knowledge the question was premature, that its resolution at that time was out of reach. So the question of how the first microscopic complexity came into being remains problematic and highly contentious."[52]

(1934) Guglielmo Marconi, Nobel Prize — Physics, 1909: "The mystery of life is certainly the most persistent problem ever placed before the mind of man. There is no doubt that from the time humanity began to think, it has occupied itself with the problem of its origin and its future — which is undoubtedly the problem of life. The inability of science to solve it is absolute."[53]

(1945) Dr. Ernst Chain, Nobel Prize — Medicine, 1945: "I have said for years that speculations about the origin of life lead to no useful purpose as even the simplest living system is far too complex to be understood in terms of the extremely primitive chemistry scientists have used in their attempts to explain the unexplainable that happened billions of years ago."[54]

(1960) Dr. Gerald Kerkut, Chairman of the School of Biochemical and Physiological Sciences and Head of the Department of Neurophysiology, University of Southampton: "The first assumption was that non-living things gave rise to living material. This is still an assumption... there is however, little evidence in favor of abiogenesis [life from non-life], and as yet we have no indication that it can be performed...it is therefore a matter of faith on the part of the biologist that abiogenesis did occur...."[55]

(1962) Dr. Harold C. Urey (mentor of Dr. Stanley Miller), Nobel Prize — Chemistry, 1934: "All of us who study the origin of life find that the more we look into it, the more we feel it is too complex to have evolved anywhere. We all believe as an *article of faith* that life evolved from dead matter on this planet. It is just that its complexity is so great, it is hard for us to imagine that it did."[56]

(1964) Dr. Cyril Ponamperumma, Chemist, Exobiologist, NASA Ames Research Center: "When we contemplate the Origin of Life, the enormity of the problem is equaled only by the complexity of the possible solutions."[57]

(1965) Dr. J. B. S. Haldane, Evolutionary Biologist, Geneticist: "If the minimal organism involves not only the code for its one or more proteins, but also twenty types of soluble RNA, one for each amino acid, and the equivalent of ribosomal RNA, our descendants may be able to make one, but we must give up the idea that such an organism could have been produced in the past, except by a similar pre-existing organism or by an agent, natural or supernatural, at least as intelligent as ourselves, and with a good deal more knowledge."[58]

(1967) Dr. David E. Green and Dr. Robert F. Goldberger, Biochemists: "[T]he macromolecule-to-cell transition is a jump of fantastic dimensions, which lies beyond the range of testable hypothesis. In this area all is conjecture. The available facts do not provide a basis for postulation that cells arose on this planet."[59]

(1970) Dr. Ernst Chain, Nobel Prize — Chemistry, 1945: "This

mechanistic concept of the phenomena of life in its infinite varieties of manifestations which purports to ascribe the origin and development of all living species, animals, plants and micro-organisms, to the haphazard blind interplay of the forces of nature in the pursuance of one aim only, namely, that for the living systems to survive, **is a typical product of the naive 19th century euphoric attitude to the potentialities of science** which spread the belief that there were no secrets of nature which could not be solved by the scientific approach given only sufficient time."[60]

(1973) Ilya Prigogine, Nobel Prize — Chemistry, 1977: "But let us have no illusions...[we are still] unable to grasp the extreme complexity of the simplest of organisms."[61]

(1974) Dr. William Thorpe, Zoologist, Professor of Animal Ethology, Cambridge University: "I think it is fair to say that all the facile speculations and discussions published during the last 10-15 years explaining the mode of origin of life have been shown to be far too simple-minded and to bear very little weight. The problem in fact seems as far from solution as it ever was."[62]

(1977) Dr. Hubert Yockey, Renowned Physicist and Information Theorist: "One must conclude that...a scenario describing the genesis of life on earth by chance and natural causes which can be accepted on the basis of fact and not faith has not yet been written."[63]

(1978) Professor Richard Dickerson, Professor of Molecular Biology, UCLA: "The evolution of genetic machinery is the step for which there are no laboratory models; hence one can speculate endlessly, unfettered by inconvenient facts."[64]

(1981) Dr. Francis Crick, Nobel Prize — Medicine, 1962: "Every time I write a paper on the Origin of Life, I determine I will never write another one, because there is too much speculation running after too few facts."[65]

"An honest man armed with all the knowledge available to us now, could only state that in some sense, the origin of life appears at the moment to be almost a miracle."[66]

(1981) Dr. Hubert Yockey: "Since science does not have the faintest idea how life on earth originated...it would only be honest to confess this to other scientists, to grantors, and to the public at large."[67]

(1982) Dr. Leslie Orgel, Biochemist and Resident Fellow — Salk Institute for Biological Studies: "Prebiotic soup is easy to obtain. We must next explain how a prebiotic soup of organic molecules, including amino acids and the organic constituents of nucleotides evolved into a self-replicating organism. While some suggestive evidence has been obtained, I must admit that attempts to reconstruct the evolutionary process are extremely tentative."[68]

(1983) Dr. Paul Davies: "The origin of life remains one of the great scientific mysteries. The central conundrum is the threshold problem. Only when organic molecules achieve a certain very high level of complexity can they be considered as 'living' in the sense that they encode a huge amount of information in a stable form and not only display the capability of storing the blueprint for replication but also the means to implement that replication. The problem is to understand how this threshold could have been crossed by ordinary physical and chemical processes without the help of some supernatural agency."[69]

(1983) Sir Fred Hoyle, Physicist, Astronomer, and Mathematician: "In short there is not a shred of objective evidence to support the hypothesis that life began in an organic soup here on the Earth."[70]

(1984) Sir Fred Hoyle: "Indeed, such a theory [Intelligent Design] is so obvious that one wonders why it is not widely accepted as being self-evident. The reasons are *psychological* rather than scientific."[71]

(1984) Chandra Wickramasinghe, mathematician, astronomer, and astrobiologist: "From my earliest training as a scientist I was very strongly brainwashed to believe that science cannot be consistent with any kind of deliberate creation. That notion has had to be very painfully shed. I am quite uncomfortable in the situation, the state of mind I now find myself in. But there is no logical way out of it; it is just not possible that life could have originated from a chemical accident." (Ibid., p. 53)

(1985) Dr. Graham Cairns-Smith, Organic Chemist and Molecular Biologist, University of Glasgow: "The singular feature is in the gap between the simplest conceivable version of organisms as we know them, and components that the Earth might reasonably have been able to generate. This gap can be seen clearly now. It is enormous... Is it any wonder that [scientists] have found the origin of life to be utterly perplexing?"[72]

(1986) Dr. Lewis Thomas, Physician, Medical Researcher, Pulitzer-Prize winning Science Writer, President of Memorial Sloan-Kettering Institute: "The events that gave rise to that first primordial cell are totally unknown, matters for guesswork and a standing challenge to scientific imagination."[73]

(1986) Dr. Andrew Scott, Biochemist, Science Writer: "In truth the mechanism of almost every major step, from chemical precursors up to the first recognizable cells, is the subject of either controversy or complete bewilderment. At the moment scientists certainly do not know how, or even if, life originated on earth from lifeless atoms."[74]

(1986) Dr. Michael Denton, Biochemist, Geneticist: "Between a living cell and the most highly ordered non-biological system such as a crystal or snowflake there is a chasm as vast and as absolute as it is possible to conceive."[75]

(1988) Dr. Klaus Dose, Biochemist: "More than thirty years of experimentation on the origin of life in the fields of chemical and molecular evolution have led to a better perception of the immensity of the problem of the origin of life on Earth rather than to its solution. At present all discussions on principal theories and experiments in the field either end in stalemate or in a confession of ignorance."[76]

(1989) Dr. Carl Woese, Microbiologist, and Dr. Gunter Wachtershauser, Chemist: "In one sense the origin of life remains what it was in the time of Darwin — one of the great unsolved riddles of science. Yet we have made progress...many of the early naïve assumptions have fallen or have fallen aside...while we do not have a solution, we now have an inkling of the magnitude of the problem."[77]

(1991) Dr. Harold P. Klein, Exobiologist, NASA: "The simplest bacterium is so damn complicated from the point of view of a chemist that it is almost impossible to imagine how it happened."[78]

(1992) Dr. Werner Arber, Biologist, Nobel Prize — Medicine 1978: "Although a biologist, I must confess that I do not understand how life came about... The most primitive cell may require at least several hundred different specific biological macro-molecules. How such already quite complex structures may have come together, remains a mystery to me."[79]

(1992) Dr. Jay Roth, Chemist, Professor Emeritus of Cell and Molecular Biology, U. of Connecticut, Storrs: "I have carefully studied molecular, biological, and chemical ideas of the origin of life and read all the books and papers I could find. Never have I found any explanation that was satisfactory to me."[80]

(1995) Dr. Stanley Miller, and Dr. Leslie Orgel, Biochemists: "It must be admitted from the beginning that we do not know how life began."[81]

(1995) Dr. Francois Jacob, Biologist, Nobel Prize — Medicine, 1965: "The simplest bacterium is already a coalition of enormous numbers of molecules. It is out of the question for all the pieces to have been formed independently in the primeval ocean, to meet by chance one fine day, and suddenly arrange themselves in such a complex system."[82]

(1995) Dr. Lynn Margulis, Biologist, National Medal of Science, 1999: "How matter in a bath of energy (or how energy in a brew of matter) first accomplished the feat of life is not known...how did the first bacterium originate? Again, no one knows."[83]

(1997) Professor Richard Dawkins: "But how did the whole process start?...nobody knows how it happened."[84]

(1998) Dr. Francois Jacob: "It goes without saying that the emergence of this RNA and the transition to a DNA world implies an impressive number of stages, each more improbable than the previous one."[85]

(1998) Dr. Laura F. Landweber and Professor Laura A. Katz, evolutionary molecular biologists: "Sherwood Chang (NASA Ames Research Center, Moffett Field, California) opened the program with the cautious reminder that any canonical scenario for the stepwise progression toward the origin of life is still just a "convenient fiction." That is, we have almost no data to support the historical transitions from chemical evolution to prebiotic monomers, polymers, replicating enzymes, and finally cells."[86]

(1998) Dr. Armand Delsemme, Astrophysicist: "The origin of life remains an immense problem and the gaps in our knowledge are countless."[87]

(1999) Dr. Robert Shapiro, Professor Emeritus of Chemistry, NYU:

"The weakest point...is our lack of understanding of the origin of life. No evidence remains that we know of to explain the steps that started life here, billions of years ago."[88]

(2000) Dr. Christopher Mckay, astrogeophysicist, NASA Ames Research Center: "The origin of life remains a scientific mystery...we do not know how life originated on the Earth."[89]

(2000) Dr. Paul Davies: "What stands out as the central unsolved puzzle in the scientific account of life is how the first microbe came to exist. Peering into life's innermost workings serves only to deepen the mystery. The living cell is the most complex system of its size known to mankind... How did something so immensely complicated, so finessed, so exquisitely clever, come into being all on its own? How can mindless molecules, capable of only pushing and pulling their immediate neighbors cooperate to form something as ingenious as a living organism? ...the problem of how and where life began is one of the great outstanding mysteries of science...many investigators feel uneasy about stating in public that the origin of life is a mystery, even though behind closed doors they freely admit they are baffled."[90]

(2000) Dr. Christian DeDuve: "According to most experts, life arose naturally by way of processes entirely explainable by the laws of physics and chemistry. However, there is no definitive proof of this statement since the origin of life is not known."[91] [That is to say, there is no proof at all.]

(2000) Nicholas Wade, Editor of *Nature* and *Science*, Science Writer for the *New York Times*: "Everything about the origin of life on earth is a mystery, and it seems the more that is known, the more acute the puzzles get... The chemistry of the first life is a nightmare to explain. No one has yet devised a plausible explanation to show how the earliest chemicals of life — thought to be RNA, or ribonucleic acid, a close relative of DNA — might have constructed themselves from the inorganic chemicals likely to have been around on the early earth. The spontaneous assembly of small RNA molecules on the primitive earth 'would have been a near miracle,' two experts in the subject helpfully declared last year... The best efforts of chemists to reconstruct molecules typical of life in the laboratory have shown only that it is a problem of fiendish difficulty. The genesis of life on earth, some time in the fiery last days of the Hadean, remains an unyielding problem."[92]

(2001) Dr. Franklin Harold, Molecular Biologist, University of Colorado: "Of all the unsolved mysteries remaining in science, the most consequential may be the origin of life... The origin of life is a stubborn problem, with no solution in sight... Biology textbooks often include a chapter on how life may have arisen from non-life, and while responsible authors do not fail to underscore the difficulties and uncertainties, readers still come away with the impression that the answer is almost within our grasp. My own reading is considerably more reserved. I suspect that the upbeat tone owes less to the advance of science than to the resurgence of primitive religiosity all around the globe, and particularly in the West. Scientists feel vulnerable to the onslaught of believer's certitudes, and so we proclaim our own... It would be agreeable to conclude this book with a cheery fanfare about science closing in slowly but surely, on the ultimate mystery; but for the time being rosy rhetoric is not yet at hand. The origin of life appears to me as incomprehensible as ever, a matter for wonder but not for explication."[93]

(2002) Dr. Ken Nealson, Microbiologist, Professor of Environmental Sciences, USC: "Nobody understands the origin of life, if they say they do they are probably trying to fool you."[94]

(2002) *Encyclopedia of Evolution*, Vol. 2, Mark Pagel, Editor in Chief, Oxford University Press, 2002: "The origin of life remains one of the most vexing issues in biology and philosophy."

(2004) Dr. E. Ben Jacob, Theoretical Physicist, Tel-Aviv University and Rice U., Y. Aharonov, Theoretical Physicist, National Medal of Science 2010, and Y. Shapira, Physicist, Professor of Microelectronics, Tel Aviv University: "Bacteria, being the first form of life on earth, had to devise ways to synthesize the complex organic molecules required for life... 3.5 billion years have passed and the existence of higher organisms depends on this unique bacterial know-how. Even for us, with all our scientific knowledge and technological advances, the ways in which bacteria solved this fundamental requirement for life are still a mystery."[95]

(2004) Dr. Andrew Knoll, Biologist, Professor of Natural History, Harvard University: "We don't know how life started on this planet. We don't know exactly when it started, we don't know under what circumstances... I don't know if [we will ever solve the problem]... I imagine my grandchildren will still be sitting around saying it's a great mystery."[96]

(2005) Dr. Stuart Kauffman, MD, Professor of Biochemistry and Biophysics, Renowned Researcher of Complex Systems: "Anyone who tells you that he or she knows how life started on the sere Earth some 3.45 billion years ago is **fool or a knave**. Nobody knows."[97]

(2005) Dr. Eugenie Scott, Biologist, Executive Director of NCSE: "Whether the proponents of [conflicting origin of life theories] finally convince their rivals as to the most plausible origin of the first replicating structures, it is clear that the origin of life is not a simple issue... If life itself is difficult to define, you can see why explaining its origin is also going to be difficult...there is not yet consensus on the sequence of events that led to living things."[98] [Talk about understatements!]

(2005) Dr. Robert Hazen, mineralogist and astrobiologist: "The epic history of life's chemical origins is woefully incomplete. Daunting gaps exist in our knowledge, and much of what we have learned is hotly debated and subject to conflicting interpretations."[99]

"What we know about the origin of life is dwarfed by what we do not know...most of the pieces are missing...and we don't [even know] what the complete picture is supposed to look like."[100]

(2007) Dr. George Whitesides, Professor of Chemistry, Harvard University: "The Origin of Life. This problem is one of the big ones in science. It begins to place life, and us, in the universe. Most chemists believe as I do, that life emerged spontaneously from mixtures of molecules in the prebiotic Earth. How? I have no idea... [Regarding the "Metabolism First" origin of life theory?] On the basis of all chemistry I know, it seems to me astonishingly improbable."[101]

(2007) Dr. Stanley L. Miller and H. James, Geochemist: "The origin of life remains one of the humankind's last great unanswered questions, as well as one of the most experimentally challenging research areas."[102]

(2007) Gregg Easterbrook, *Wired Magazine*: "What creates life out of the inanimate compounds that make up living things? No one knows. How were the first organisms assembled? Nature hasn't given us the slightest hint. If anything, the mystery has deepened over time."[103]

(2008) Dr. Robert Shapiro: "I'm always running out of metaphors to try and explain what the difficulty is. But suppose you took Scrabble sets, or any word game sets, blocks with letters containing every language on Earth and you heap them together, and then you took a scoop and you scooped into that heap, and you flung it out on the lawn there and the letters fell into a line which contained the words, "to be or not to be that is the question." That is roughly the odds of an RNA molecule appearing on the earth."[104]

(2008) Dr. J. Craig Venter, synthetic biologist and physiologist (in a remark to Dr. Robert Shapiro about the absence of any plausible naturalistic theory of the origin of life): "The theory behind theory is that you come up with truly testable ideas. Otherwise it's no different than faith. It might as well be a religion if there's no evidence for it."[105]

(2008) Dr. Freeman Dyson, Theoretical Physicist, Mathematician: "First of all I wanted to talk a bit about origin of life...That has been a hobby of mine. We're all equally ignorant as far as I can see. That's why somebody like me can pretend to be an expert."[106]

(2008) Dr. Alexandre Meinesz, Professor of Biology: "In sum, all of these vestiges of very ancient life provide no precise clues to the place, time, and mechanism of the genesis of the first living organisms. Therefore the currently popular idea that life probably arose in warm, subsurface waters along a mid-ocean ridge...is a hypothesis without any scientific basis...no transition process between inorganic matter and bacteria has been found in nature...it is a fact that at the beginning of the third millennia, we cannot yet describe and illustrate the processes and the stages in the genesis of bacteria. The exact time and place of the spontaneous generation of the first bacteria remain unknown...we must humbly recognize that...the birth of life on Earth is only an unsupported hypothesis; all research trying to confirm it is at an impasse. It is just an idea...that has been taught. This idea has become a dogma."[107]

(2009) Dr. Jack Szostak, Biochemist, Professor of Genetics, Nobel Prize — Medicine 2009: "Understanding how life emerged on Earth is one of the greatest challenges facing modern chemistry."[108]

(2009) Dr. Chris Wills, Professor Emeritus, Biological Sciences, UC San Diego: "The biggest gap in evolutionary theory remains the origin of life itself."[109]

(2009) Dr. Jeffrey Bada (Professor of Marine Chemistry) and Dr. Antonio Lazcano (Evolutionary Biologist): "Although there have been considerable advances in the understanding of chemical processes that may have taken place before the emergence of the first living entities, life's beginnings are still shrouded in mystery... [H]ow the transition from the non-living to the living took place is still unknown."[110]

(2009) Dr. Ken Miller, Professor of Biology, Brown University: "However, the most profound unsolved problem in biology is the origin of life itself."[111]

(2009) Nature.com blogs (Nature Publishing Group), Anna Kushnir (March 9, 2009): "The Origins of Life Initiative and the Harvard Alumni Association hosted a day long symposium at Harvard's Science Center entitled 'The Future of Life' focused on discussing...what is life, and how did it begin? [Among the speakers were] J. Craig Venter...Jack Szostak...George Whitesides...George Church... It may be difficult to believe, but there was a common theme to this seeming cacophony of scientific expertise and discovery. The theme was 'we just don't know.' No one knows how life began or even how to define 'life' if you want to get all philosophical about it... Underneath it all, it was refreshing to hear a bunch of really smart folks say 'we don't know.' It was humbling and put things in a grandiose perspective. No one knows how we all got to be here, but the researchers in the Origins of Life Initiative and beyond are trying to find out."

(2009) Drs. Stochel, Brindel, Macyk, Stasicka, Szacilowski, Microbiologists: "Most of the biochemical processes found within all the living organisms are well understood at the molecular level, whereas the origin of life remains one of the most vexing issues in chemistry, biology, and philosophy... Numerous theories tackle the origin of life, but there is no direct evidence supporting any of them."[112]

(2009) Dr. Jack Szostak and Dr. Alonso Ricardo: "The actual nature of the first organisms and the exact circumstances of the origin of life may be forever lost to science... One of the most difficult and interesting mysteries surrounding the origin of life is exactly how the genetic material could have formed."[113]

(2009) Dr. Seth Shostak, Astronomer at SETI Institute: "Finding clues to life's earliest moments on Earth is tougher than overcooked road kill. The trail...is cold...so it's hard to work from direct evidence...this list of theories as to how biology might have begun on Earth, abbreviated in consideration of the reader's finite life span, should convince you that we still haven't figured out what really happened."[114]

(2010) Dr. Milton Wainwright, Microbiologist: "So much has been written about the Origin of Life that it might seem that little else needs to be said. Despite the lack of conclusive or convincing evidence it is generally accepted that life originated on Earth from simple chemicals... Are we getting any closer to an understanding of the origin of life?... The reality is that, despite the egos of some, the existence of life remains a mystery. It is not merely that biology is scratching the surface of this enigma; the reality is that we have yet to *see* the surface!"[115]

(2010) The Origin of Life Gordon Research Conference (Jan. 10-15, 2010, Galveston, Texas): "The origin and early development of life whether specifically on earth, or possibly elsewhere in the universe, remains one of the great unsolved scientific problems."[116]

(2010) Dr. Gerald Joyce, Chemist, Scripps Research Inst., and Dr. Michael Robertson: "The concept of an RNA World has been a milestone in the scientific study of life's origins. **While this concept does not explain how life originated**, it has helped to guide scientific thinking and has served to focus experimental efforts."[117]

(2010) Dr. Paul Davies: "How [did life begin]? We haven't a clue."[118]

(2010) Dr. Freeman Dyson: "The origin of life is the deepest mystery in the whole of science. Many books and learned papers have been written about it, but it remains a mystery. There is an enormous gap between the simplest living cell and the most complicated naturally occurring mixture of nonliving chemicals. We have no idea when and how and where this gap was crossed."[119]

(2010) Dr. Timothy Kusky, Professor of Earth Sciences: "Complex organic molecules including amino acids do not constitute life. After the simple amino acids form, it is no easy task to combine them into larger molecules and complex molecules necessary for life... Somehow, in the early Precambrian, life emerged from these complex organic molecules and simple amino acids, but the origin of life remains one of life's biggest mysteries."[120]

(2011) Dr. Christian DeDuve: "While much has been learned, it is clear that we are still nowhere near explaining the origin of life."[121]

(2011) John Horgan, Senior Writer, *Scientific American*: "Dennis Overbye just wrote a status report for the New York Times on research into life's origin, based on a conference on the topic at Arizona State University. Geologists, chemists, astronomers, and biologists are as stumped as ever by the riddle of life."[122]

(2011) Dr. Paul Davies: "When I was a student in London in the swinging sixties...the prevailing view at the time was summed up by Francis Crick who said that life seems almost a miracle, so many are the conditions necessary for it to get going. What he meant by this was that it's entirely possible that life on earth was a bizarre freak event, an aberration unique in the entire universe... Today you scarcely open a newspaper without reading that scientists think that the universe is teeming with life. What, you may wonder has changed, do we now know how life began so that we can confidently say, yes, it's everywhere? Well, we don't know how life began...we don't know the mechanism that turned non-life into life...we have...many conjectures but we don't know what happened."[123]

(2011) Dr. Darrell Falk, Biologist & founder of the Biologos website: "From all that we know about the state of Earth three to four billion years ago and what we know about the complexity of the building blocks of life — DNA, RNA, amino acids, sugars — no entirely plausible hypothesis for the spontaneous origin of life has been found...the origin of life is simply a particularly compelling example of an unsolved mystery we would like to understand."[124]

(2011) Dr. Eugene Koonin, Microbiologist: "The origin of life field is a failure...certainly this is due not to a lack of...effort, but to the extraordinary intrinsic difficulty and complexity of the problem. A succession of exceedingly unlikely steps is essential for the origin of life... these make the final outcome seem almost like a miracle."[125]

(2012) The Origin of Life Gordon Research Conference: "The origin of life on Earth, and its possible existence elsewhere in the universe, offer some of science's greatest unsolved problems."[126]

(2012) Professor James Tour, Professor of Chemistry, Computer Science, and Mechanical Engineering and Materials Science, Rice University (one of the top ten most cited chemists in the world): "Let me tell you what goes on in the back rooms of science — with National Academy members, with Nobel Prize winners. I have sat with them, and when I get them alone, not in public — because it's a scary thing, if you say what I just said — I say, 'Do you understand all of this, where all of this came from, and how this happens?' Every time that I have sat with people who are synthetic chemists, who understand this, they go, 'Uh-uh. Nope.' These people are just so far off, on how to believe this stuff came together. I've sat with National Academy members, with Nobel Prize winners. Sometimes I will say, 'Do you understand this?' And if they're afraid to say 'Yes,' they say nothing. They just stare at me, because they can't sincerely do it.

"I was once brought in by the Dean of the Department, many years ago, and he was a chemist. He was kind of concerned about some things. I said, 'Let me ask you something. You're a chemist. Do you understand this? How do you get DNA without a cell membrane? And how do you get a cell membrane without a DNA? And how does all this come together from this piece of jelly?' [His response was:] 'We have no idea, we have no idea.' I said, 'Isn't it interesting that you, the Dean of science, and I, the chemistry professor, can talk about this quietly in your office, but we can't go out there and talk about this?'"[127]

(2013) The Origin of Life Gordon Research Conference: "Originating Life in the Lab, 7:30–9:30 pm — Ok. Maybe we cannot solve the historical question: How did life actually arise on Earth. Can we originate some of our own life by "intelligent design?"[128]

(2014) Johnny Bontemps, *Astrobiology* Magazine: "The story of life's origin is one of the great unsolved mysteries of science. The puzzle boils down to bridging the gaps between two worlds — chemistry and biology. We know how molecules behave and we know how cells work. But we still don't know how a soup of lifeless molecules could have given rise to the first living cells. 'It's a really tough problem,' says Sara Walker, an astrobiologist at the University of Arizona."[129]

(2014) Dr. Addy Pross, Professor of Chemistry, Ben-Gurion University, Israel: "Despite the profound advances in molecular biology in the past half century, we still do not understand what life is, how it relates to the inanimate world, and how it emerged. True, over the past half century considerable effort has been directed into attempts to solve these fundamental issues, but the gates to the Promised Land seem as distant as ever. Like a mirage in the desert, just as the palm trees signaling the oasis seemingly materialize shimmering on the horizon, they fade away yet again... Yet paradoxically our digging deeper and deeper into the mechanisms of life did not seem to lead any closer to being able to address [the] question [of]how did life emerge?...we are still lacking a theory of life, a theory that will enable us to understand what life is and how it emerged..."[130]

(2015) Dr. Sukhendu B. Dev, Geneticist, Chief Scientist at Gene Delivery and Expression Sciences, San Diego, CA, visiting fellow: Welcome Unit for the History of Medicine, Oxford University: "However, all these examples I have given show how complex and challenging the problem of origin of life is...although [it] may never be solved..."[131]

(2015) Real Clear Science: "When biologists get together to discuss the nagging mysteries in their diverse field, there's always that elephant in the room: How did life spring up from non-life? But, according to highly regarded cancer researcher Robert Weinberg, it's an elephant that most biologists ignore, or at least discreetly avoid. 'Origin of life is not something people work on that much because it's so far away from resolution.'"[132]

In the 156 years between 1859–2015, not only has Origin of Life research made zero progress, it has actually moved *backwards* by many orders of magnitude. In 1859 scientists had no idea at all how life began; in 2015 scientists still have no clue how life began, but due to the many breakthroughs in genetics and microbiology, their understanding of the extraordinary difficulty of the problem has grown exponentially. As Dr. Lee Hartwell (Nobel Prize in Medicine, 2001) put it at the aforementioned ASU Origins Conference: "With respect to the origin [of life], I find the more we learn about cells the more complex they seem; they are just incredibly complex things, and to go from what we can see today and try to reason where it came from, I think is really impossible."[133]

At this point we are ready to draw the obvious conclusions regarding Evolution, Design, Origin of Life, and the existence of a Creator of life:

- Remember, the skeptic has never denied the very high levels of functional complexity and design that are apparent in the living world.
- The skeptics and non-theists have never denied the validity of the Argument from Design. Instead, they have claimed that the Argument from Design does not apply to living organisms because they are an exception to the rule. All the functional complexity, sophistication, and design in the living world can be explained by the theory of Darwinian Evolution.
- We have demonstrated that this line of reasoning is not only misleading, but false. Discussions about evolution have absolutely no relevance to the question of the existence of a Creator of life and do not in any way address the questions raised by the Argument from Design. Attempting to explain the origin of functional complexity and specified information necessary for life through evolution is tautological; it explains the origins of functional complexity and specified information in living systems by invoking the functional complexity and specified information of living systems.
- The Argument from Design predicts that any attempt to explain the origins of functional complexity and specified information as being the result of some unguided undirected process is hopeless and will end in failure. I defy any scientist to present me with an exception to the rule.
- The area of scientific investigation that is most directly confronted and challenged by the Argument from Design is not evolution but rather the Origin of Life. The Argument from Design would label

any attempt to explain the origin of the functional complexity and specified information of a living cell through a naturalistic, unguided process as being *nonsense of a high order.* This phrase, from whence the title of the book was taken, was coined by one of the great scientific minds of the twentieth century, Sir Fred Hoyle. The full citation reads as follows:

> Imagine 10^{50} blind persons each with a scrambled Rubik's cube, and try to conceive of the chance of them all simultaneously arriving at the solved form. You then have the chance of arriving by random shuffling, of just one of the many biopolymers on which life depends. The notion that not only the biopolymers but the operating program of a living cell could be arrived at by chance in a primordial organic soup here on the Earth is evidently nonsense of a high order.[134]

- We have *predicted correctly* that any investigation into the absurd notion of the emergence of a living bacterial cell through a naturalistic, undirected process would not only be fruitless but would leave investigators baffled and confused.

Where Does This Leave Us?

It leaves us all alone with a nanotechnology-filled molecular machine that exhibits stunning levels of sophistication, stunning levels of specified digitally encoded information, stunning levels of functional complexity, and no explanation for its origin. In fact, it leaves us right back where we started from: that my suit *itself* is the evidence for the existence of the tailor who created the suit and *life itself* is the evidence for the existence of the Creator of life. Just as we are absolutely certain that the existence of the suit implies the existence of the tailor, so too are we absolutely certain that the incredible machinery of life itself implies the existence of the Creator of life.

Now that we have arrived at this point, when it has become clear that the non-believer cannot hide anymore behind Darwinian Evolution, when it turns out that the entire foundation of scientific credibility for atheism is nothing more than a mirage, when the potent challenge of Origin of Life is met with scientific shoulder-shrugging; what *is* the response of the atheist/skeptic/non-theist?

The objections to the arguments I have presented generally fall into one of two categories:

1. **Non-Scientific**: The notion of a Creator outside of the material universe is **(a)** not in the category of science and need not be considered, and in any case, **(b)** scientists are busy working on the problem.
2. **Philosophical**: There are two main philosophical attacks on the logic of the Argument from Design: **(a)** "The Argument from Ignorance" (also known as the "God of the Gaps Argument") and **(b)** "Who Created the Creator, Who Designed the Designer?"

In the following chapters we will thoroughly confront and address each of these objections.

End Notes

1 John Horgan, *The End of Science: Facing the Limits of Knowledge in the Twilight of the Scientific Age* (New York, NY: Broadway Books, 1996), p. 138.

2 Eugenie Scott, *Evolution vs. Creationism — an Introduction* (University of California Press, 2005, Paperback Edition), p. 27.

3 Leonard Susskind, *The Black Hole War: My Battle with Stephen Hawking to Make the World Safe for Quantum Mechanics* (New York: Little, Brown and Company, 2008), p. 128.

4 Ibid., p. 129.

5 Robert Shapiro, *Origins: A Skeptics Guide to the Creation of Life on Earth* (Summit Books, 1986), pp. 168–169.

6 Richard Dawkins, "The Illusion of Design," *Natural History*, Nov. 2005.

7 Richard Dawkins, *The Blind Watchmaker: Why the Evidence of Evolution Reveals a Universe without Design* (New York: W.W. Norton & Company, Ltd., 1996), p. 4.

8 Jerry Coyne, *Why Evolution Is True*, p. 1.

9 Lawrence Krauss, "A Universe without Purpose," *L.A. Times*, April 1, 2012.

10 Francis Crick, *What Mad Pursuit: A Personal View of Scientific Discovery*, p. 138.

11 "About Us" page on the Wyss Institute website, http://wyss.harvard.edu/viewpage/about-us/about-us

12 Shapiro, *Origins*, pp. 125–126.

13 Robert Shapiro, "A Simpler Origin for Life," *Scientific American*, Feb. 12, 2007, http://www.scientificamerican.com/article/a-simpler-origin-for-life/

14 Richard Dawkins, *River Out of Eden: A Darwinian View of Life* (New York: Basic Books, 1995),

p. 19.

15 http://www.slate.com/articles/health_and science/new_scientist/2013/12/richard_dawkins_interview_science_religion_irrationality_group_selection.html

16 "Life," Encyclopedia Britannica Macropaedia, 15th ed. (Chicago: Encyclopedia Britannica, 1974), pp. 893–894.

17 "Passing the Baton of Life — From Schrodinger to Venter," *New Scientist*, July 13, 2012, http://www.newscientist.com/blogs/culturelab/2012/07/passing-the-baton-of-life---from-schrodinger-to-venter.html)

18 https://www.sciencemag.org/content/337/6102/1628.abstract

19 http://www.nature.com/news/dna-data-storage-breaks-records-1.11194

20 Bill Gates, *The Road Ahead* (London: Penguin Books, 1996), p. 228.

21 Dr. Paul Davies, *The Fifth Miracle: The Search for the Origin and Meaning of Life* (New York: Simon and Schuster, 2000), p. 115.

22 "Bacteria as Multicellular Organisms," *Scientific American* Vol. 258, no. 6 (June, 1988): p. 82.

23 "The Origin of Life on Earth," *Scientific American*, August 19, 2009, p. 54.

24 "The Cell as a Collection of Protein Machines: Preparing the Next Generation of Molecular Biologists," *Cell* (Feb. 8, 1998): p. 291.

25 David Suzuki and Holly Dressel, *From Naked Ape to Superspecies*, rev. ed. (Vancouver: Greystone Books, 2004), p. 172.

26 Davies, *The Fifth Miracle*, p. 29.

27 Ibid., pp. 97–99.

28 Dr. Michael Denton, *Evolution: a Theory in Crisis* (Chevy Chase, MD.: Adler & Adler, 1986), p. 234, 328.

29 Dr. Michael Denton, *Nature's Destiny: How the Laws of Biology Reveal Purpose in the Universe,* (New York: Free Press, 1998), p. 212–213.

30 G. C. Smith, *Seven Clues to the Origin of Life* (Cambridge University Press, 1985), p. 14.

31 Dr. Paul Davies, *Cosmic Blueprint: New Discoveries in Nature's Creative Ability to Order the Universe* (West Conshohocken, PA: Templeton Foundation Press, 2004), p. 115.

32 Alexandre Meinesz, *How Life Began* (University of Chicago Press, 2008), p. 32.

33 Richard E. Dickerson, "Chemical Evolution and the Origin of Life," *Scientific American*, Vol. 239, No. 3 (September 1978): p. 77.

34 *Encyclopedia of Evolution*, Volume 2 (Oxford University Press, 2002).

35 Stuart Kauffman, *At Home in the Universe: The Search for the Laws of Self-Organization and Complexity* (Oxford University Press, 1995), p. 42.

36 Thomas Nagle, "The Fear of Religion," *The New Republic*, October 2006, http://www.tnr.com/article/the-fear-religion

37 Davies, *Cosmic Blueprint*, p. 115.

38 Dr. Addy Pross, *What is Life?: How Chemistry Becomes Biology* (Oxford University Press, 2012), p. 8.

39 www.pandasthumb.org/archives/2005/10/robert-shapiro.html

40 Horgan, *The End of Science*, pp.140–141.

41 Denton, *Evolution: A Theory in Crisis*, p. 249.

42 Eugene Koonin, *The Logic of Chance: The Nature and Origin of Biological Evolution* (Upper Saddle River, NJ: FT Press, 2011), p. 391.

43 Francis Crick, *Life Itself: Its Origin and Nature* (New York: Simon & Schuster, 1981), p. 88.

44 Smith, *Seven Clues to the Origin of Life*, p. 8.

45 Ibid., p. 30.

46 Davies, *The Fifth Miracle*, p. 98.

47 Ibid., p. 82.

48 Ibid., p. 93.

49 Euan Nisbet, *The Young Earth*, (Kluwer Academic Publishers, 1987), http://www2.glos.ac.uk/GDN/origins/life/index.htm

50 http://www.nytimes.com/2011/02/22/science/22origins.html?_r=1&ref=science https://origins.asu.edu/events/great-debate-what-life

51 http://blogs.scientificamerican.com/cross-check/2011/02/28/pssst-dont-tell-the-creation-ists-but-scientists-dont-have-a-clue-how-life-began/

52 Pross, *What is Life?*, p. 8.

53 From a scientific address by Marconi to the International Congress of Electro-Radio Biology, Venice, Italy, September 10, 1934.

54 R. W. Clark, *The Life of Ernst Chain: Penicillin and Beyond* (London: Weidenfeld and Nicolson, 1985), p. 148.

55 Gerald Kerkut, "Implications of Evolution," *International Series of Monographs on Pure and Applied Biology — Division: Zoology, Volume 4* (Pergamon Press, 1960), p. 150.

56 *Christian Science Monitor, January 4, 1962.*

57 "The Origin of Life in the Universe," Space Science Education Conference (1964), Los Angeles, CA, https://archive.org/stream/nasa_techdoc_19650011877/19650011877#page/n0/mode/2up

58 J. B. S. Haldane, "Data Needed for a Blueprint of the First Organism," in S. W. Fox, ed., "The Origins of Prebiological Systems and of Their Molecular Matrices," *Proceedings of a Conference Conducted at Wakulla Springs, Florida, October 27–30, 1963* (New York, NY: Academic Press, 1965), p. 12

59 David Ezra Green and Robert F. Goldberger, *Molecular Insights into the Living Process*, (New York: Academic Press, 1967), p. 407.

60 E. Chain, *Social Responsibility and the Scientist in Modern Western Society* (London: The Council of Christians and Jews, 1970), pp. 24–25.

61 Ilya Prigogine, *Impact of Science on Society (1973).*

62 Dr. William Thorpe, "Reductionism in Biology," in Francisco Ayala and Theodosius Dobzhansky, eds., *Studies in the Philosophy of Biology: Reduction and Related Problems* (Berkeley, CA: University of California Press, 1974), p. 116.

63 H. P. Yockey, "A Calculation of the Probability of Spontaneous Biogenesis by Information Theory," *Journal of Theoretical Biology*, 1977.

64 Richard Dickerson, "Chemical Evolution and the Origin of Life," *Scientific American, September 1978, p. 77.*

65 Francis Crick, *Life Itself, p. 153.*

66 Ibid., *p. 88.*

67 H. P. Yockey, "Self-Organization Origin of Life, Scenarios and Information Theory," *Journal of Theoretical Biology*, Vol. 91, no. 1, (July 7, 1981): p. 13.

68 Leslie Orgel, "Darwinism at the Very Beginning of Life," *New Scientist (April 1982): p. 150.*

69 Dr. Paul Davies, *God and the New Physics* (New York: Simon & Schuster, 1983), p. 68.

70 Sir Fred Hoyle, *The Intelligent Universe* (New York: Holt, Rinehart & Winston), p. 23.

71 Hoyle, *Evolution From Space (New York: Simon & Shuster, 1984), p. 148.*

72 Smith, *Seven Clues*, p. 4, 15.

73 Found in foreword to Robert M. Pool, ed., *The Incredible Machine* (Washington, DC: National Geographic Book Service, 1986), p. 7.

74 Andrew Scott, *The Creation of Life: Past, Future, Alien* (Oxford, England: Basil Blackwell, 1986), p. 111.

75 Denton, *Evolution: A Theory in Crisis*, p. 233.

76 Klaus Dose, "The Origin of Life: More Questions Than Answers," *Interdisciplinary Science*

Reviews, vol. 13, no. 4, (1988): p. 348.

77 "Origin of Life" in *Paleobiology: A Synthesis, Briggs and Crowther, eds. (Oxford: Blackwell Scientific Publications, 1989).*

78 Quoted in John Horgan, "In the Beginning...", *Scientific American*, Feb. 1991.

79 Werner Arber, "The Existence of a Creator Represents a Satisfactory Solution," in H. Margenau, H. and R. A. Varghese (eds.), *Cosmos, Bios, Theos: Scientists Reflect on Science, God, and the Origins of the Universe, Life, and Homo sapiens* (La Salle, IL: Open Court), pp. 141–143.

80 Jay Roth, Ibid., p. 199

81 Dr. Lynn Margulis, *What Is Life?* (Los Angeles, CA: University of California Press, 1995), p. 57.

82 Ibid.

83 Ibid., p 58.

84 Richard Dawkins, *Climbing Mt. Improbable* (New York: Norton Paperback Edition, 1997), p. 282.

85 Francois Jacob, *Of Flies, Mice and Men*, trans. Giselle Weiss (Harvard University Press, 1998), p. 21.

86 "Evolution: Lost Worlds," *Trends in Ecology and Evolution*, Vol. 13 (March 1998), pp. 93–94.

87 Armand Delsemme, *Our Cosmic Origins: From the Big Bang to the Emergence of Life and Intelligence* (New York: Cambridge University Press, 1998), p. 160.

88 Robert Shapiro, *Planetary Dreams: The Quest to Discover Life beyond Earth* (New York: John Wiley & Sons, Inc., 1999), p. 26.

89 Christopher Mckay, "Astrobiology: The Search for Life Beyond the Earth," in *Many Worlds* (Radnor, PA: Templeton Foundation Press, 2000), p. 49.

90 Davies, *The Fifth Miracle*, pp. 17, 27, 29–30.

91 Christian DeDuve, "Lessons of Life" in *Many Worlds* (Templeton Foundation Press), p. 5.

92 "Life's Origins Get Murkier and Messier; Genetic Analysis Yields Intimations of a Primordial Commune," *New York Times*, June 13, 2000.

93 Franklin Harold, *The Way of the Cell: Molecules, Organisms, and the Order of Life (New York: Oxford University Press, 2001), p. 236, 251.*

94 Robert Roy Britt, "The Search for the Scum of the Universe," Space.com (May 21, 2002), http://www.space.com/scienceastronomy/astronomy/odds_of_et_020521-1.html)

95 E. Ben Jacob, Y. Aharonov, Y. Shapira, "Bacteria Harnessing Complexity," *Biofilms* (Cambridge University Press, 2004).

96 "How Did Life Begin?" *NOVA*, July 1, 2004, http://www.pbs.org/wgbh/nova/beta/evolution/how-did-life-begin.html

97 Quoted in Dr. Robert Hazen, *Genesis: The Scientific Quest for Life's Origin* (Washington DC: Joseph Henry Press, 2005), p. 241.

98 Scott, *Evolution vs. Creationism*, pp. 25–26.

99 Hazen, *Genesis*, xiv.

100 Ibid., p. 241.

101 "Revolution in Chemistry" — Priestley Medalist George Whitesides' Address, *Chemical and Engineering News*, Vol 85, No.13 (March 26, 2007): pp. 12–17, http://pubs.acs.org/cen/coverstory/85/8513cover1.html

102 "Prebiotic Chemistry on the Primitive Earth," in Isidore Rigoutsos and Gregroy Stephanopoulos, eds., *Systems Biology Volume 1: Genomics* (New York: Oxford University Press, 2007), p. 3.

103 Gregg Easterbrook, "Where Did Life Come From?" *Wired*, February 2007, p. 108.

104 Robert Shapiro, "Life: What a Concept!", *Edge*, 2008, p. 84, http://www.edge.org/documents/life/Life.pdf

105 Ibid., p. 101.

106 Ibid., p. 11.

107 Alexandre Meinesz, *How Life Began*, pp. 30–33.

108 As cited by Dr. Ada Yonath, 2009 Nobel Laureate — Chemistry, http://scienceblogs.com/terrasig/2009/12/ada_yonath_interview.php
109 Chris Willis, "Evolution: The Next 200 Years," *New Scientist*, Jan. 28, 2009, http://www.newscientist.com/article/mg20126932.600-evolution-the-next-200-years.html?full=true
110 Jeffrey Bada, "The Origin of Life," in Michael Ruse and Joseph Travis, eds., *Evolution: The First Four Billion Years* (Cambridge, MA: Belknap Press, 2009), p. 72.
111 Ken Miller, "Evolution: the Next 200 Years" *ad loc.*
112 G. Stochel, *et al*, *Bioinorganic Photochemistry* (UK: John Wiley and Sons, Ltd.), p. 109.
113 Alonso Ricardo and Jack Szostak, "The Origin of Life on Earth," *Scientific American*, September 2009, http://www.scientificamerican.com/article.cfm?id=origin-of-life-on-earth
114 Seth Shostak, *Confessions of an Alien Hunter* (National Geographic Society, 2009), pp. 82–89.
115 Milton Wainwright, "Musings on the Origin of Life and Panspermia," *Journal of Cosmology*, Vol. 5 (Jan. 30, 2010).
116 From the description of the conference on the Gordon Research Conferences website, http://www.grc.org/programs.aspx?year=2010&program=origin)
117 Dr. Michael Robertson and Dr. Gerald Joyce, "The Origins of the RNA World" (Cold Spring Harbor Perspectives in Biology, Cold Spring Harbor Laboratory Press, April 28, 2010), p. 18.
118 http://www.youtube.com/watch?v=jGaId-E0BF4
119 Freeman J. Dyson, *A Many Colored Glass: Reflections on the Place of Life in the Universe* (Charlottesville, VA: University of Virginia Press, 2010), p. 104.
120 *Encyclopedia of Earth and Space Science* (New York: Facts on File, 2010), p. 384.
121 Christian DeDuve, "Mysteries of Life: Is There 'Something Else'?" in Bruce L. Gordon and William A. Dembski, eds., *The Nature of Nature: Examining the Role of Naturalism in Science* (Wilmington, DE: ISI Books, 2011), p. 349.
122 John Horgan, *"Psst, Don't Tell the Creationists But Scientists Don't Have a Clue How Life Began," Scientific American,* Feb. 28, 2011.
123 From his presentation at the 2011 ASU Origins Conference.
124 http://biologos.org/questions/the-origin-of-life
125 Eugene Koonin, *The Logic of Chance: The Nature and Origin of Biological Evolution* (Upper Saddle River, NJ: FT Press, 2011), p. 391.
126 Gordon Research Conferences – Origin of Life Conference, January 2012, Galveston, Texas, http://www.grc.org/programs.aspx?year=2012&program=originlife
127 James Tour at Georgia Tech, 2012, https://www.youtube.com/watch?v=PZrxTH-UUdI
128 Description of a session to be held at the Gordon Research Conferences – Origins of Life Conference, January 2014, Galveston, Texas, http://www.grc.org/programs.aspx?year=2014&program=origins
129 Johnny Bontemps, "From Soup to Cells: Measuring the Emergence of Life," *Astrobiology Magazine*, March 7, 2014, http://www.astrobio.net/topic/origins/origin-and-evolution-of-life/from-soup-to-cells-measuring-the-emergence-of-life/
130 Addy Pross, *What is Life?*, pp. 8, 38, 43.
131 Dr. Sukhendu B. Dev, "Unsolved Problems in Biology — The State of Current Thinking," *Progress in Physics and Molecular Biology, (March 2015).*
132 "Intriguing Mysteries of Life," *Real Clear Science, May 13, 2015.*
133 "What is Life? The Great Debate," ASU Origins *Conference* https://www.youtube.com/watch?v=xIHMnD2FDeY
134 Sir Fred Hoyle, "The Big Bang in Astronomy," *New Scientist (Nov. 19, 1981): p. 527.*

Chapter 4

Scientists Are Very Human

Objection #1: An Intelligent Designer Is Not in the Category of Science

Here is a small sampling of those who raise this objection:

- Union of Concerned Scientists (Ucsusa.org): From an article entitled Science, Evolution, and Intelligent Design — "Section 4: Why Intelligent Design is Not Science."[1]
- University of California Museum of Paleontology: From the "Understanding Science" section: Intelligent Design: Is it Scientific? — "Intelligent Design is very different from Science..."[2]
- Dr. Robert Kirshner, president of the American Astronomical Society: "Intelligent Design isn't even part of Science."[3] "Since "Intelligent Design" is not science, it does not belong in the science curriculum of the nation's primary and secondary schools."[4]
- American Society for Biochemistry and Molecular Biology: "Intelligent Design is not a theory in the scientific sense...the concept has no place in a science classroom and should not be taught there."[5]
- United States National Academy of Sciences: "Intelligent Design and other claims of supernatural intervention in the origin of life...are not science."[6]

Before we directly address the objection voiced above, the entire issue of scientific exploration and scientists themselves must be placed in their proper perspective. Many lay people have somehow accepted the notion that scientists are qualitatively different than the rest of humanity. The impression is that the scientist, through some fortuitous mutation, has leapfrogged to the top of the human ladder; *Homo scientistus*, one step above the rest of us lowly *Homo sapiens*. In his book, *The Monkey Business*, renowned evolutionary paleontologist Niles Eldredge addresses this fanciful view of the scientific world:

> Many scientists really do seem to believe they have a special access to the truth. They call press conferences to trumpet marvelous new discoveries. They compete hard for awards and prizes. And they expect to be believed — by their peers and, especially, by the public at large. Throwing down scientific thunderbolts from Olympian heights, scientists come across as authoritarian truth givers whose word must be taken unquestioned. That all the evidence shows the behavior of scientists clearly to be no different from the ways in which other people behave is somehow overlooked in all this.[7]

Eldredge writes elsewhere: "In the competitive fray that is science, data forging, plagiarism, and all manner of base but utterly human failings make a mockery of the counter image of detached objectivity."[8]

Dr. Stephen J. Gould seconds Eldredge; in his view, the notion that scientists are detached robots operating on pure logic is simply hokum:

> Our ways of learning about the world are strongly influenced by the social preconceptions and biased modes of thinking that each scientist must apply to any problem. The stereotype of a fully rational and objective "scientific method" with individual scientists as logical (and interchangeable) robots is self-serving mythology.[9]

"Back Off, I'm a Scientist!"

(Bill Murray, in the film **Ghostbusters***)*

Scientists as a whole are no better or worse than the practitioners of any other trade or profession. There are scientists who are scrupulously honest and of impeccable integrity, and there are some who are petty, vindictive, and underhanded. It may be true that there is a system in place to keep scientists on the straight and narrow; for example, if a scientist wishes to publish an article in a respected journal, it must be reviewed by his peers. However, there is also a regulatory system in place to oversee the activities of bank executives, but that doesn't mean that there aren't bankers who will cook the books when they find an opportunity, will find other ways to fudge data to their own advantage, and will even engage in outright thievery if they think they can get away with it. Scientists want honor, recognition, and job security just like everyone else.

Furthermore, peer review of articles is not a magical formula that creates mythical levels of integrity and perfection in science and scientists. Biologist Gunter Blobel, in a news conference given just after he was awarded the Nobel Prize in Medicine in 1999, said that the main problem one encounters in one's research is "when your grants and papers are rejected because some stupid reviewer rejected them for dogmatic adherence to old ideas." According to the *New York Times*, these comments "drew thunderous applause from the hundreds of sympathetic colleagues and younger scientists in the auditorium."[10] All is not perfect in Science Land and there are clearly many scientists out there who are not happy campers.

Important scientific work has not uncommonly been initially rejected by peer-reviewed journals. Martin Enserink, in a 2001 article in *Science* entitled: "Peer Review and Quality: A Dubious Connection?" observed: "Mention 'peer review' and almost every scientist will regale you with stories about referees submitting nasty comments, sitting on a manuscript forever, or rejecting a paper only to repeat the study and steal the glory."[11]

Dr. Randy Schekman, awarded the 2013 Nobel Prize for medicine, angrily announced that his lab was boycotting top science journals *Nature*, *Cell*, and *Science* because they are "distorting the scientific process" and represent a "tyranny" that must be broken. Schekman claimed that pressure to publish in these "luxury journals" encourages researchers to cut corners and engage in trendy areas of science rather than doing more important work.[12] Competition for government research grants is intense and one of the criteria used is how often a researcher or his team's papers are published in prestigious journals — whether or not the research is particularly significant. Nobel Laureate Randy Schekman is *not* a happy camper.

Rosalyn Yalow, a medical physicist, described how her Nobel Prize-winning paper was received by the journals: "In 1955 we submitted the paper to *Science*... The paper was held there for eight months before it was reviewed. It was finally rejected. We submitted it to the *Journal of Clinical Investigations*, which also rejected it."[13]

In an article for *Twentieth Century Physics*, a book commissioned by the American Physical Society (the professional organization for US physicists) to describe the great achievements of twentieth century physics, Mitchell J. Feigenbaum, the inventor of chaos theory, described the reception that his revolutionary papers on chaos theory received:

> This has been my full experience. Papers on established subjects are immediately accepted. Every novel paper of mine, without exception, has been rejected by the refereeing process. The reader can easily gather that I regard this entire process as a false guardian and wastefully dishonest.[14]

The title says it all in physicist Juan Miguel Campanario's 2009 article: "Rejecting and Resisting Nobel Class Discoveries: Accounts by Nobel Laureates." From the abstract:

> "I review and discuss instances in which nineteen future Nobel Laureates encountered resistance on the part of the scientific community towards their discoveries, and instances in which twenty-four future Nobel Laureates encountered resistance on the part of scientific journal editors or referees to manuscripts that dealt with discoveries that later would earn them the Nobel Prize."[15]

Campanario continues:

> The history of science is dotted with stories documenting how

> many important discoveries were initially resisted or ignored by fellow scientists. Some important discoveries...did not fit the common paradigms...these discoveries were often rejected...in other instances novel theories or discoveries collided with dominant paradigms in science, and were resisted and scorned...authors of very innovative papers are criticized and often face stonewalling from their peers...
>
> Despite the deluge of documented cases there has been a relative lack of interest on the part of sociologists, philosophers, and science historians in investigating the important topic of scientists' resistance to scientific discovery. It is naturally embarrassing for the scientific community to acknowledge that many important discoveries were neglected, rejected, or utterly ignored.[16]

The obvious question is: How many important discoveries or important new ideas have been completely lost because the researcher either gave up or decided that if there is so much resistance to the paper, he or she must have been mistaken? My purpose here is not to cast aspersions on the scientific community as a whole, belittle scientific accomplishments, nor in any way to undermine respect for the scientific endeavor. It simply is to remind people that scientists are very human...just like the rest of us.

Scientists have exhibited neither a higher nor lower moral caliber than any other group of people. In Nazi Germany, Imperial Japan, and the former Soviet Union, as well as other places in the world, scientists have been responsible for mind-numbing atrocities and the production of horrific weapons of destruction. Scientists are tempted by greed, jealousy, lust, pride, and envy, no more and no less than any other group. Some scientists act on their prejudices and look to promote their agenda whenever they have the opportunity. Furthermore, by their own admission, scientists suffer at times from an acute case of "groupthink."

Dr. Bernard Nathanson (1926–2011), one of the founders of the National Association to Repeal Abortion Laws (NARAL), described in his autobiography how he used his scientific credentials to spread false propaganda and flat out lies in order to further his organizations pro-abortion agenda.[17] Scientific "consensus" must often be taken with a grain of salt.

"I sometimes think scientists really don't notice their colleagues have flaws. But in my experience scientists are very human people, which means that some are troubled, deceitful, petty, or vain."[18] (Dr. Michael Crichton, author of *Jurassic Park*)

With this perspective, it should therefore not be surprising to discover that right along with the many truly stellar accomplishments that have emerged from the world of science, there is an accompanying history of stupendous blunders and unjustified attitudes and assumptions that scientists have made in their attempts to understand the natural world. All human foibles including arrogance, wishful thinking, narrow-mindedness, blind obedience to "standard teachings," pure stubbornness, and even political affiliations have contributed to some of the serious errors that scientists have made over the centuries. Just a few examples:

- Dr. Robin Warren, an Australian pathologist, was scoffed at when he suggested that some stomach ulcers were caused not by stress, as was the standard teaching, but rather by bacteria residing in the intestines. His collaborator on this project, Dr. Barry Marshall, was told by the chief gastroenterologist at one of Australia's major hospitals that Dr. Warren was "the crackpot downstairs trying to prove that bacteria cause gastritis." In a *Time* interview in 2005, he recalled the scorn that was heaped on him for daring to challenge the conventional scientific wisdom by suggesting that bacteria caused peptic ulcers: "It was pretty savage." He was even denied access to tissue samples with which to conduct his research. In February 1983, their paper was rejected by the Gastroenterological Society of Australia. By 1984, their work had demonstrated that most peptic ulcers *were* caused by bacteria that somehow managed to survive the acidity of the human stomach. Millions of sufferers who were once given palliatives for a chronic condition could now be cured with antibiotics. Warren and Marshall were awarded the Nobel Prize for Medicine in 2005.[19]
- America's first Nobel Laureate in Physics (1907), Albert Michelson, declared the following in an 1894 speech given at the University of

Chicago: "The most important fundamental laws and facts of physical science have all been discovered, and these are now so firmly established that the possibility of their ever being supplemented in consequences of new discoveries is exceedingly remote...Our future discoveries must be looked for in the sixth place of decimals."[20] How ironic in light of the fact that Michelson's own experiments helped disprove the long held theory of the luminiferous ether. Little did he know that a young physicist working in a patent office in Switzerland by the name of Albert Einstein was about to revolutionize our entire understanding of the physical world with his theory of relativity.

- When renowned physicist Max Planck was a twenty-year-old graduate student, one of his professors, Phillip von Jolly, advised him against becoming a physicist. He told him that after the discovery of the two laws of thermodynamics, all that was left to do was to tie up loose ends. This was how the *physics community* saw things as the twentieth century approached. Just a few years later, Planck formulated the theory of the quantum, which helped usher in revolutionary developments associated with quantum mechanics and relativity.[21]
- In the late eighteenth century, astronomer Sir Frederick William Herschel rocked the world of astronomy by announcing the discovery of a seventh planet that came to be known as Uranus. Since antiquity, it was assumed that there were only six planets (Earth, Mars, Venus, Mercury, Jupiter, and Saturn), those that are visible to the naked eye. A *Scientific American* article stated that "the idea that our solar system harbored a whole other world...captivated astronomers... [They found] that the planet had actually been seen twenty times prior to 1781, including as early as 1690, but misidentified as a star." Herschel's predecessors could literally not see what they saw because they had been conditioned to believe there were only five other planets besides Earth. It had never been challenged or questioned. It took an amateur astronomer like Herschel to break through the established "known" facts and think outside the box.[22]
- Professor Dan Shechtman of the Technion Institute in Israel was also ridiculed by the entire scientific community for his discoveries

regarding the nature of quasicrystals that ran contrary to the "accepted" scientific paradigms. In his own words: "For a long time it was me against [the] world. I was a subject of ridicule...the leader of the opposition...was the two-time Nobel Laureate Linus Pauling, one of the most famous scientists in the world...for years, until his last day, he fought against quasi-periodicity in crystals." Pauling once said, "There is no such thing as quasicrystals, only quasi-scientists." At one point the head of Shechtman's research group told him to "go back and read the textbook" and asked him to leave for "bringing disgrace" on the team. In 2011, Shechtman was awarded the Nobel Prize in Chemistry for his research.[23]

- Cosmologists make absolutist declarations about the size and age of the universe while at the same time hypothesizing that somewhere in the neighborhood of **95%**(!) of our universe consists of *dark matter and dark energy* regarding whose nature we are absolutely clueless. In fact, the names "dark matter and energy" are simply another way of saying "we have no idea what it is." The reason why physicists hypothesize the existence of dark matter and energy is because certain observed phenomena make no sense in light of the *visible and measurable* amount of matter and energy in the universe.[24] (Isn't a little bit of humility in order here?)

The list goes on and on, but I will add a particularly heart-rending example. Dr. Phillip Ignaz Semmelweis (1818–1865) was a Hungarian physician who was appointed as chief resident of the First Obstetrics Clinic of the Vienna General Hospital in 1846. At the time, an alarming number of women giving birth ended up dying from what was called "puerperal fever." Strangely enough, the mortality rates were significantly lower among women who gave birth *outside* of the hospital. This was one of the clues that eventually led Semmelweis to discover the cure for the disease. He wrote that the number of women dying in his clinic "made life so miserable that life seemed worthless." What turned out to be the "cure" for puerperal fever is shocking to those unfamiliar with the story.

Semmelweis discovered that the occurrence of the disease could be effectively reduced to zero if doctors would simply *wash their hands* with a chlorine/lime solution when they went from *dissecting cadavers* to delivering babies. Remember, this was well before Louis Pasteur and the development of germ theory. These findings ran against the conventional scientific wisdom that diseases spread in the form of "bad air," also known as miasmas. His groundbreaking idea that cleanliness was crucial in preventing the spread of disease ran contrary to established medical/scientific understanding.

One would have thought that the results themselves — the startling reduction in the mortality rate when his protocols were followed — would have taken the medical profession by storm. What Semmelweis could not have foreseen and what he did not count on was the arrogance and stubbornness of the medical/scientific community when faced with a paradigm shift. Not only were his findings rejected but he was ridiculed and ostracized by the members of his own profession. According to the Wikipedia entry, "Some doctors were offended by the suggestion that they should wash their hands, feeling their social status as gentlemen was inconsistent with the idea that their hands could be unclean... Semmelweis was outraged by the indifference of the medical profession and began writing open and increasingly angry letters to prominent European obstetricians, at times denouncing them as irresponsible murderers." He eventually was dismissed from his position and tragically died in an insane asylum. He was not fully vindicated until well after his death. This story should be an object lesson for those who find themselves overawed by the men who wear white lab coats.

Scientific Expertise Is Not Transferable

Scientific expertise is not in any way transferable to other disciplines. A Noble Prize-winning chemist may have no more expertise in world history than an average high school student; in fact he may have less. A Nobel Prize-winning physicist does not necessarily have any more insight into the wisdom of living a happy life and success in marriage and raising children than the plumber who comes to fix his sink. It goes further than that. He may have no more expertise in philosophy, psychology, morality, and theology than any particular Orthodox rabbi. I defer to the physicist exclusively on the subject of *physics*.

When a brilliant and highly accomplished scientist expresses opinions or draws conclusions in areas that are outside of his expertise, or when it is clear that he is being motivated by pre-conceived notions or a personal agenda, it would be a flagrant assault on the entire concept of intellectual integrity to grant him special dispensation due to his standing in the scientific world. It is high time we jettison naïve notions about the infallibility, objectivity, and integrity of scientists. Scientific knowledge most definitely changes with time, however, human nature with all its failings does not. As Nobel Prize-winning physicist Max Planck informed us: "A new scientific truth does not triumph by convincing its opponents and making them see the light, but rather because its opponents eventually die, and a new generations grows up that is familiar with it."[25] Strong words indeed that every truth-seeking individual should contemplate.

In this chapter, I will bring example after example to demonstrate that Origin of Life researchers routinely draw conclusions in areas outside of their expertise and are consistently motivated by pre-conceived notions and/or personal agenda. In fact, it is my contention that among too many of these scientists, it has become standard operating procedure. With this, then, we are ready to begin directly addressing the objection that Intelligent Design or a Creator of life need not be considered because they not in the category of science.

If You Are Only Prepared to Consider One Possibility, then There Is Only One Possibility

Imagine an east-west highway with a rest stop standing in the median between the two roads. There is an exit ramp from the eastbound highway leading to the rest stop and an identical ramp from the westbound highway. Inasmuch as there is no other access to the rest stop, there are only two possible ways for a vehicle to enter: from the east or from the west. Two men standing in the rest stop notice a car pull up without either having seen from which direction it came. The following dialogue ensues:

Man #1: Which direction do you think that car came from?

Man #2: Well, *barring the possibility* that it came from the west, it must have come from the east.

Man #1: What kind of ridiculous answer is that? There are only two possibilities to begin with. All you have told me is that if we only consider one possibility, then there is only one possibility. That is like saying one equals one. That type of statement is known as a *tautology*. It is pointless, meaningless, and gives me no new information.

Man #2: Well, what did you want from me when you asked the question?

Man #1: I wanted to know if you had any observations or empirical evidence that would indicate which direction it came from. In other words, do you have any evidence that would indicate which of the two possibilities is true?

Dr. Robert Hazen, a highly accomplished scientist who has devoted many years to investigating the origins of life on Earth, in his acclaimed book, *Genesis: The Scientific Quest for Life's Origin*, informs us of the different possible explanations for the origin of life: "How did life arise...? *Barring Divine intervention*, life must have emerged by a natural process — one fully consistent with the laws of chemistry and physics.[26]

What point is Dr. Hazen making when he states that "barring Divine intervention, life must have emerged by a natural process"? Is this analysis and conclusion based on carefully conducted laboratory experiments, other empirical evidence, or solid scientific reasoning? Hazen has essentially told us that there are two possibilities to explain the Origin of Life: (a) Divine intervention, i.e., an intelligent designer outside of the material universe, and (b) natural processes fully consistent with the laws of physics and chemistry. One would think that the next step would be to explain how the rational truth seeker should go about investigating and deciding between these two possibilities. Incredibly though, he offers us nothing more than a tautology, analogous to the above-stated example of the two travelers in the rest stop:

Man #1: Dr. Hazen, in your opinion, was the Origin of Life the result of (a) Divine intervention or (b) natural processes fully consistent with the laws of physics or chemistry?

Hazen: Well, "*barring Divine intervention*, life must have emerged through a natural process."

Man #1: What kind of ridiculous response is that? You have informed us that there are only two possibilities to begin with. If you eliminate one possibility, then of course only one possibility remains!

Simply stated, if we arbitrarily decide to only consider one of the possibilities, then there is only one possibility. Hazen's statement is as meaningful, informative, logical, and *scientific* as the following: "*Barring naturalistic processes*, life must have emerged by Divine intervention." The relevant question that must be answered is: *Why* are you barring Divine intervention?

In fact, Hazen does offer an explanation as to *why* he has chosen to ignore the possibility of a supernatural creator:

> Scientists **believe** [a very scientific term] in a universe ordered by natural laws; they resort to the power of observations, experiments, and theoretical reasoning to discover those laws... Scientists **surmise** [even more scientific!] that life arose on the blasted, primitive Earth from the most basic of raw materials: air, water, rock. Life emerged nearly four billion years ago by natural processes completely in accord with the laws of chemistry and physics...[27]

Dictionary.com: Surmise — "To think or infer without certain or strong evidence; conjecture, guess."

How did scientists conjecture or guess ("surmise") that life emerged by "natural processes completely in accord with the laws of chemistry and physics?" Because "scientists *believe* in a universe ordered [only] by natural laws." Can this be considered "scientific reasoning" by any stretch of the imagination?! There is no question that this statement is based on a leap of faith, for in the very next sentence (and any number of other places in the book), he explicitly informs us that it is not based on evidence:

> ...Yet details of that transforming origin even pose mysteries as deep as any facing Science. How did non-living chemicals become alive?...The epic history of life's chemical origins is woefully incomplete.[28]
>
> ...What we know about the origin of life is dwarfed by what we

> don't know. It's as if we were trying to assemble a giant jigsaw puzzle. A few pieces clump together here and there, but most of the pieces are missing and we don't even have the box to see what the complete picture is supposed to look like.[29]

He even cites the rather scathing and caustic remark by prominent Origin of Life researcher, Dr. Stuart Kauffman: "Anyone who tells you that he or she knows how life started on the sere* Earth some 3.45 billion years ago is a fool or knave."[30] Dr. Hazen then tells us that he is even prepared to consider the possibility of blind luck, despite the probability calculations that have relegated to the realm of complete and utter irrationality the idea that the first living organisms arose by pure chance:

> It is possible, of course, that life arose through an improbable sequence of many chemical reactions. If so then living worlds will be rare in the universe and laboratory attempts to understand the origin process will be doomed to frustration. An unlikely sequence of unknown steps cannot be reproduced in any plausible experimental program.[31]

I have some important news for Dr. Hazen. Divine intervention in the origin of life is either *true* or it is *not true*. Divine intervention does not magically appear just because I or anyone else believes in it, but neither does it magically disappear just because someone is averse to considering its truth. It does not cease to be an issue or a possibility because it is dismissed with a shallow offhand remark.

A Summary of Dr. Hazen's "Scientific" Approach to Origin of Life

- An arbitrary, *a priori* rejection of the possibility of an intelligent Designer or Divine intervention.
- After this arbitrary rejection, a tautological declaration that life must have emerged through natural processes, the only remaining possibility.

---

* Arid, parched.

- Explaining this position by using non-scientific — or even anti-scientific — terms like *belief* or *surmise*, all the while camouflaging it as a scientific approach by using one's credentials and expertise and relying on the "consensus" of other scientists who have done the same thing to boost its credibility.
- Stating explicitly that there is no scientific evidence to back up this position and that the Origin of Life is a complete mystery; but this fact has no influence on our conclusions since we have already declared that there is only one possibility that can be considered.
- When all else fails, the fallback position is never to consider the option of intelligent intervention but rather to rely on *astoundingly improbable blind luck*!

It may be very hard for many of the readers to believe that an accomplished scientist could write such foolishness, but then it is also hard to believe that scientists savagely attacked a researcher for suggesting that bacteria cause peptic ulcers and the criminal indifference of the medical/scientific world when it was suggested that they could save countless lives by simply washing their hands before delivering babies. When agendas and comfortably held worldviews are at stake, there is no limit to the amount of foolishness that can be generated in their defense, by scientist and non-scientist alike. We will soon identify the agenda and worldview at stake in this issue. Before we do that, however, it is important to understand that Dr. Hazen's approach reflects a general pattern in the Origin of Life field.

Dr. Hazen Is Not Alone in His Non-scientific Approach

A highly respected philosopher of science confirms that Dr. Hazen is not alone in his decidedly non-scientific approach to Origin of Life. In fact the above-stated non/anti-scientific paradigms define the approach of nearly every scientist engaged in the research of life's origins. Researchers either take a leap of faith and arbitrarily reject the possibility of intelligent design or believe in miraculous strokes of luck.

Dr. Iris Fry, a non-believing philosopher of science at Tel-Aviv University and the Technion Institute in Haifa (and an opponent of Intelligent

Design theory), in a seminal paper entitled "Are the Different Hypotheses on the Emergence of Life as Different as They Seem?" writes that there are essentially two approaches taken by researchers in the Origin of Life field. It is important to note that she describes both not as *scientific* positions (because as we've pointed out, scientists have no real clue as to how life arose), but as *philosophical* positions. The first is what she calls "the continuity thesis":

> This paper calls attention to a **philosophical presupposition** coined here "the continuity thesis"...this presupposition, a necessary condition for any scientific investigation of the origin of life problem has two components. First, it contends that there is no unbridgeable gap between inorganic matter and life. Second, it regards the emergence of life as a highly probable process.[32]

I again call the readers attention to the very non-scientific language used in the description of those who espouse this position: "[I]t *contends* that there is no unbridgeable gap between inorganic matter and life... it *regards* the emergence of life as a highly probable process." These are terms that reflect personal feelings and opinions, not scientific or empirically-based facts. Fry makes it very clear that this "philosophical presupposition" is (a) not based on evidence and (b) *is non-falsifiable*.

It is extremely important at this juncture to explain the implications of non-falsifiability. The concept of "falsifiability" is crucial in scientific research; it is a critical factor in determining whether a hypothesis can even be considered scientific at all. From an article on the subject: "**Falsifiability** or **refutability** of a statement, hypothesis, or theory is an inherent possibility to prove it to be false. A statement is called **falsifiable** if it is possible to conceive of an observation or an argument which proves the statement in question to be false...what is un-falsifiable is classified as unscientific, and the practice of declaring an un-falsifiable theory to be scientifically true is pseudoscience."[33] We now continue with Dr. Fry's description of the continuity thesis:

> The various principles of continuity might indeed push forward the experimental investigation of the emergence of life; as such

> they do represent the heuristic [educational] advantage of the continuity thesis. However, **the decision to adopt the continuity thesis is a philosophical one...and this decision does not depend on the success of a specific experimental program, nor can it be revoked on the basis of its failure.**[34]

Translation: The *a priori* assumption of the existence of an (as yet unknown) set of coherent, ordered physical laws and processes that inevitably lead from non-life to life is a crucial pre-condition in conducting origin of life research. In that sense, the "continuity thesis" pushes forward experimental investigation of the matter. However, the *decision* to adopt such a position is not based on any experimental evidence or a particular experimental program, and that failure of any particular experiment or avenue of investigation will not invalidate this decision. In short, it is a position that is based on a *decision*, not evidence, and cannot be falsified by experimentation.

What do we generally call a decision that is (a) not based on evidence, (b) is adopted because it advances an agenda (in this case justifying investigation of a naturalistic origin of life), and (c) cannot be falsified through experimentation? It is self-apparent that Dr. Fry has mistakenly and unjustifiably dignified this position by describing it as *philosophical.* In fact, such a position is nothing more than an *article of faith*, as stated explicitly by Nobel Laureate, Dr. Harold Urey: "All of us who study the Origin of Life find that the more we look into it, the more we feel it is too complex to have evolved anywhere. We all believe as an *article of faith* that life evolved from dead matter on this planet. It is just that its complexity is so great, it is hard for us to imagine that it did."[35] Simply put, the "continuity thesis" is non/anti-scientific.

She then goes on to describe the "rival" hypothesis, espoused by, among others, Nobel Prize winners Francis Crick and Jacque Monod and famed biologist Ernst Mayr:

> In addition I identify the rivals of the [continuity] thesis within the scientific community — "**the almost miracle camp.**"...This camp regards the emergence of life as involving highly improbable events... The basic philosophical assumption underlying the

> "almost miracle" notion becomes apparent, once we learn that for Crick, the emergence of life was "a happy accident."[36]

Upon careful examination, however, it becomes clear that Dr. Fry has also seriously erred by describing the "almost miracle" camp as a *philosophical* position. Please bear with me as I elaborate. Dr. Jacque Monod, "one of the most pronounced representatives of this position," claims that the origin of life is not so much a "problem" as a "veritable enigma." In order to understand this enigma, a little bit of background information is in order.

Protein-Machines in Living Cells

All proteins in living cells, which are essential for the survival of the organism, consist of various arrangements of twenty different amino acids. Think of the proteins as words that are composed of a twenty-letter alphabet. These proteins consist of hundreds, and in many cases thousands, of amino acids that must be linked together in a *very precise* order for them to function properly. The challenge that a living cell faces in accomplishing this task is exactly the same as one who has been given the task of stringing together individual letters to form coherent and sophisticated words, sentences, and paragraphs. The nonsense combinations greatly outnumber functional combinations. Let us let another Nobel Prize-winning member of the "almost miracle" camp, Dr. Francis Crick, explain the dilemma:

> All the cell needs do is to string together the amino acids...in the correct order... Here we need only ask how many possible [combinations of] proteins are there? If a particular amino acid sequence was selected by chance, how rare of an event would that be? This is an easy exercise in combinatorials.* Suppose the chain is about two hundred amino acids long; this is, if anything, rather less than the average length of proteins of all types. Since we have just twenty possibilities at each place, the number of possibilities is twenty multiplied by itself some two hundred times. This is...

* I.e., the statistical analysis of the number of different possible combinations.

> approximately 10^{260}, that is a one followed by 260 zeroes! The number is quite beyond our everyday comprehension...the number of fundamental particles (atoms, speaking loosely) in the entire visible universe...is estimated to be 10^{80}... [It is] quite paltry by comparison to 10^{260}. Moreover we have only considered a [protein] chain of a rather modest length. Had we considered longer ones as well, the figure would have been even more immense.[37]

As Dr. Monod explains in his classic work *Chance and Necessity*, there are no chemical or physical laws that determine any particular order of amino acids to build the first proteins; all amino acids can essentially link together equally well.[38] The vast majority of the almost limitless number of combinations of amino acids will produce meaningless and non-functional strings of chemicals, in the same way that the monkey banging away at a typewriter will produce gibberish, because meaningful combinations of letters are vastly outnumbered by nonsense combinations.

How functional proteins formed *in the first place* is only the beginning of this mathematical/probability nightmare. How does the living cell then "know" the proper sequence in which to link the amino acids to produce functional proteins? The simplified answer is that the correct order is coded in the sequence of four different nucleobases — **A**-adenine, **T**-thymine, **G**-guanine, **C**-cytosine — in their vertical arrangement on the double-helix of the DNA molecule. Each combination of three nucleobases, called a "codon" (for example, **ATG**) codes for a particular amino acid. The coding for a protein that consists of two hundred amino acids would correspond then to six hundred nucleobases (200 amino acids multiplied by three corresponding nucleobases for each one) on the DNA molecule. The double helix of the DNA molecule is first unwound, its coded information read and copied in the form of a specific type of RNA (called messenger–RNA) and then fed into a highly complex protein-machine (also composed of amino acids) called a ribosome. With the help of other types of RNA and molecular protein-machinery, the ribosome reads the code and in assembly-line fashion links together the appropriate amino acids to produce the proteins needed by the cell.

The enigma now increases exponentially because just like the linkages of amino acids, there are no physical or chemical laws that make any particular arrangement of nucleobases more likely than any other. In their vertical arrangement on the DNA molecule, all can link together equally well. Remember, the nucleobases must also be in the correct order to code for the production of the various functional proteins. For a protein like hemoglobin, which consists of six hundred amino acids, *eighteen hundred* corresponding nucleobases must be in the proper order.

Dr. Stephen Meyer, in his seminal work on Intelligent Design theory *Signature in the Cell*, presented the perfect analogy to illustrate the problem: The physical properties that cause magnetic letters to adhere to a refrigerator door are identical, analogous to the identical physical properties that cause nucleobases to link together. However, there are absolutely no physical or chemical properties that determine the *order* in which the letters are arranged on the door.[39] Therefore, if you walk into your kitchen and the letters on the fridge read: "I am coming home late dear, but I left meat loaf in the freezer that you can heat up" (i.e., specified information), we know beyond any reasonable doubt that they are the product of conscious, intelligent intent, *not some chemical or physical necessity.* How then did the hundreds of thousands, actually millions, of nucleobases "know" the proper order in which to arrange themselves on the strands of DNA in order to code for the production of functional proteins? We now understand why Francis Crick described the production of proteins in a living cell as "a miracle of molecular construction."[40]

The IBM Supercomputer

The following description of an IBM project to build a supercomputer will flesh out for the reader the staggering magnitude of the problem Monod and Crick are describing. In October of 1999, IBM announced that it was launching a $100 million research initiative to build a computer that could perform a quadrillion computations per second [a quadrillion is one followed by fifteen zeros, or a thousand trillion]. The project was dubbed Blue Gene. The purpose of the computer was "to help researchers understand how proteins are created, knowledge that could lead to a better understanding of diseases and uncover possible cures." An article on CNET News went on to describe in more detail what tasks the computer would perform:

> Blue Gene will take on the problem of protein folding, the biochemical process by which complex molecules are constructed by instructions carried in the DNA. As proteins are assembled from components called amino acids, the long strand of molecules twists and folds in a three-dimensional bundle, leaving some active sites protruding from the protein to react with the environment.
> "How exactly the protein will fold up is governed by basic rules of how atoms attract and repel each other," [Paul Horn, senior vice-president of IBM research] said. But the size of proteins, often with thousands of atoms, makes predicting that arrangement a very difficult task. Hemoglobin — also known as the red blood cells that carry oxygen throughout the body — is made of 600 amino acids, for example. Blue Gene's final product, due in four or five years, will be able to "fold" a protein made of 300 amino acids, Horn said. "**But that job will take an entire year of full-time computing.**"[41]

Once the molecular machinery of a cell assembles a particular protein, within minutes it folds into its precise three-dimensional shape without which the protein will not function. This means that the microscopic cellular protein factory enables the precision folding process of proteins to function somewhere in the neighborhood of 200,000–500,000 times faster than a computer making a thousand trillion computations per second!

There is an even deeper problem to contend with regarding the proposition of a naturalistic emergence of life. One of the most serious challenges in copying large amounts of information, which happens endlessly in each living cell, is that errors creep in. With each generation of replication the errors increase exponentially, resulting in what is called "error catastrophe." In other words, errors quickly multiply to the point where the information is useless. Therefore, the bacterial cell must, and does, contain a highly sophisticated error correction system to ensure the integrity of the replication process and the copying of information to synthesize proteins and all the other materials necessary for the survival of the organism.

Where does the complex molecular machinery needed to retrieve and translate information and correct copying errors come from? How are the amazing molecular machines that perform these functions constructed? (It's worth noting that we are talking about devices that are measured in the billionths of a meter.) The answer is simple: the information and instructions required to build them, including the molecular machinery that performs the actual building process, are all contained in the coded sequences of nucleobases in the DNA. Much like information stored in a computer hard drive, the information in DNA is useless unless it can be retrieved and translated with copying integrity ensured. *The machinery for retrieval, translation, and error correction cannot be produced without a previous error-free retrieval and translation of the information contained in the DNA*. We can now appreciate more fully the previously cited statement by Dr. Graham Cairns-Smith:

> There seems to be a more fundamental difficulty. Any conceivable kind of organism would have to contain messages of some sort and equipment for reading and reprinting the messages; any conceivable organism would thus have seem to have to be packed with machinery and as such **need a miracle (or something)** for the first of its kind to have appeared.[42]

Noble Prize-winning biochemist Jacque Monod put it this way:

> But the major problem is the origin of the genetic code and of its translation mechanism. Indeed, instead of a problem it ought rather to be called a riddle. The code is meaningless unless translated. The modern cell's translating machinery consists of at least fifty macromolecular components which are themselves coded in DNA: the code cannot be translated otherwise than by products of translation. It is the modern expression of *omne vivum ex ovo* [all life comes from life]. When and how did this circle become closed? It is exceedingly difficult to imagine.[43]

I leave it to the reader to contemplate this ultimate version of the "which came first; the chicken or egg?" conundrum. Let us continue now with the conclusions drawn by the "almost miracle camp" of researchers as described by Dr. Fry.

Inasmuch as neither the order of amino acids in functional proteins nor the arrangement of the genetic coding of the nucleobases in DNA are determined by operative physical or chemical laws, the only option left for Dr. Monod to explain the origin of this incredible system is chance. He acknowledges that the random probability of such a system coming into being is "virtually zero."* Luckily for us, he writes, "our number came up in the Monte Carlo game." Monod himself admitted how distasteful to him as a scientist his "casino" conclusion was, for he acknowledged that "science can neither say nor do anything about a unique occurrence. It can only consider events which form a class whose *a priori* probability, however faint, is definite."[44]

Renowned philosopher of science, Karl Popper, agrees with Monod that life could only emerge from inanimate matter by an extremely improbable combination of chance circumstances and admits that the Origin of Life becomes "an impenetrable barrier to science." In the same vein, Ernst Mayr asserts that "a full realization of the near impossibility of an origin of life brings home the point how improbable this event was." Finally Francis Crick, who together with James Watson was awarded a Nobel Prize for discovering the structure of DNA in 1953, writes in his book *Life Itself* that an honest man armed with all the knowledge and evidence available to us now could only state "in some sense the origin of life appears at the moment to be almost a miracle, so many are the conditions which would have had to been satisfied get it going."[45]

In other words, Mayr, Crick, Popper, and Monod do not describe the Origin of Life as being "almost a miracle" based on *philosophical considerations*. Their conclusions are reasoned, logical statements based on *examination of the available evidence*. The results of their research are that the continuity thesis is not supported in any way by scientific evidence;

* Compare to the earlier cited evaluation of Sir Fred Hoyle, who described the notion of such a system coming into existence randomly as being "nonsense of a high order."

in fact it flies in the face of the evidence. They explicitly admit that Origin of Life is a unique event and therefore outside the realm of science, even "an impenetrable barrier to science."

In short, one of the positions taken by Origin of Life researchers as described by Dr. Fry — that life *inevitably* must emerge from non-life under the right conditions — is an article of faith. (In fact Dr. Eors Szathmary, professor of biology at Eotvos Lorand University, Budapest, has quite appropriately dubbed this position as "the gospel of inevitability."[46]) The other position, that the emergence of life is "almost a miracle" is a reasoned conclusion based on clear scientific investigation.

I would pose a simple question to Dr. Monod and his "secular miracle" believing colleagues: If "science can neither say nor do anything about a unique occurrence," then the entire question of the origin of life is outside of the realm of science, why then is Dr. Monod even offering an opinion on what happened? His scientific credentials have no standing or significance anymore in determining how life began and therefore his statement that "our number came up in the Monte Carlo game" is nothing more than a wild guess on his part. A wild guess on the part of a scientist is no more or less valid than the wild guess of a cab driver.

This is now a problem not for scientists, but for statisticians. The math tells us that the probability of this event occurring as a result of unguided, naturalistic forces is "virtually zero," this means that the only alternative — intelligent design — is "virtually certain." The arrangement of the nucleobases, a four-letter alphabet which, in combinations of three, code for hundreds of amino acids to be linked together to form the functional proteins, is directly analogous to the arrangement of the magnetic letters on a fridge to form sophisticated, intelligible information. The probability of an infant, monkey, or some other unguided, random process producing an arrangement of fifty letters on a refrigerator door that spelled out an intelligible message is "virtually zero," and in fact, as we all know, is essentially impossible. The obvious, undeniable conclusion is that they were arranged by conscious intelligent activity. Why wouldn't Monod and his colleagues concede that obvious point and at least *consider the possibility* that a creator/designer of life is at work here?

The answer is very clear. Just like Dr. Hazen, these scientists have arbitrarily and tautologically decided that the only permissible option for the origin of life is an unguided naturalistic process. They tenaciously cling to this position no matter how absurd and *anti-scientific* it may seem and even if rational thinking must go out the window to keep it intact.

In fact, even *science fiction* is preferable to Divine creation. In 1973, Francis Crick and Leslie Orgel published an article in the journal *Icarus* suggesting that life may have arrived on Earth through a process called *Directed Panspermia*; that is to say, life was sent here from outer space by an advanced extra-terrestrial civilization. I wonder if Crick and Orgel also made the pilgrimage to Roswell, New Mexico, where UFO enthusiasts are certain that the US government has secretly imprisoned men from outer space for decades. How many times *did* Francis Crick watch the classic 1956 B-movie *Invasion of the Body Snatchers*?

Can This Be Called Science?

It is very difficult to understand why men and women of science, who ostensibly pride themselves on rational and reasoned thinking, and on drawing conclusions based on experimental evidence, would choose to rely on wild conjecture, blind luck, or feel forced to accept an *article of faith* as the basis for scientific investigation. All of these approaches can be easily avoided and it will be quite revealing when we discover why scientists choose not to do so.

Contrary to what Dr. Fry wrote, it is rather obvious that adopting the "philosophical presupposition" of the continuity thesis is most definitely *not* a "necessary condition for any scientific investigation of the origin of life." There is a very simple way for researchers to continue their investigation into the origin of life on Earth and at the same time not demean their dignity as scientists and rational thinkers by adopting non-falsifiable articles of faith. This point is so important it warrants a full explication.

Dr. Stuart Kaufman writes the following in his book, *At Home in the Universe*:

> Indeed we may never recover the actual **historical sequence** of

> molecular events that led to the first self-reproducing, evolving molecular systems to flower forth more than three million millennia ago. But if the **historical pathway** should forever remain hidden, we can still develop bodies of theory and experiment to show how life might have realistically crystallized, rooted, then covered our globe. Yet the caveat: nobody knows.[48]

Origin of Life on this planet is clearly in the category of an *historical event*. Dr. Kauffman, accurately uses the terms "historical sequence" and "historical pathway" when describing the events that led to the flowering of life on Earth. *Something* happened some 3.7 billion years ago that resulted in living organisms swarming all over our planet. Nobody was around to witness these events. It is up to us to examine the evidence and deduce what happened. The following analogy will illustrate the proper approach that should be taken by scientists in this field.

What Is the Proper Way for an Historian to Investigate an Historical Event?

The distinguished historian, Barbara Tuchman (1912–1989), wrote a best-selling, Pulitzer Prize-winning historical work, *The Guns of August*, describing the events that led up to the outbreak of World War I. Let us assume that she was unquestionably an authoritative historical voice on that period of history. Imagine that Barbara Tuchman, in an address to the faculty of the history department of Harvard University, proposed that Kaiser Wilhelm made a secret trip to the United States in 1913 in an attempt to get American support for Germany if war broke out in Europe. The following exchange then ensued:

Harvard Historian: "Dr. Tuchman, that is quite a startling proposal. None of us have ever heard of such an event taking place. To the best of our knowledge, Kaiser Wilhelm never left Germany for the entire year before the outbreak of WWI. What evidence do you have for this claim?"

Tuchman: "I have no evidence at all that Kaiser Wilhelm actually made this trip, but I *believe, surmise, speculate*, and *accept as an historical presupposition* that such an event occurred, and since I am a world recognized authority on the history of World War I, it stands to reason that I will eventually uncover evidence that confirms my presupposition."

Harvard Historian: "Dr. Tuchman, your qualifications as an historian are impeccable. No one is more qualified than you to investigate *if* such an event actually took place or the *plausibility* of such an event taking place; but the *historical truth* is that Kaiser Wilhelm either made the secret visit or he did not; if it did not actually happen, all the academic credentials in the world cannot make it true. To claim that such an event actually happened based solely on your previously established authority as an historian is nothing less than *utter nonsense.*"

The *historical truth* is that either life began as the result of a naturalistic unguided process or it did not. The *historical truth* is that life was the result of an intelligent act of creation or it was not. The *scientific truth* is that a naturalistic origin of life is either plausible or it is not.

Scientists like Dr. Kauffman are eminently qualified to investigate *if* life arose through a series of naturalistic unguided steps; they are eminently qualified to investigate *if* a naturalistic origin of life is *plausible* or not. However, all the scientific accomplishments in the world cannot make an historical event a reality *if it never actually happened.* All the PhDs in the universe cannot make an historical event plausible, *if in fact it is not.* Because scientists were able to use all the genius and ingenuity at their disposal to discover a vaccine for polio or to unlock the power of the atom and create thermonuclear weapons does not give them the magical ability to create historical/scientific realities. It becomes obvious, then, that to claim life arose through an unguided process with no evidence to support such a claim, other than the fact that one has a PhD in Chemistry, is *utter nonsense!*

Why then does Kauffman seem unable to distance himself from this folly by qualifying his statement with the addition of a single word: "We can still develop bodies of theory and experiment to investigate *if* life crystallized, rooted, and then covered our globe [through an unguided naturalistic process]"? Why can't Dr. Fry distance herself from this same folly by writing the following: "First, Origin of Life researchers are investigating *if* there is a bridgeable gap between inorganic matter and life.

Second, they are investigating *if* the emergence of life is a highly probable process."

Dr. Paul Davies also seems helpless to extricate himself from the very same trap: "I should like to say that the scientific attempt to explain the origin of life proceeds from the *assumption* that whatever it was that happened was a natural process... Scientists *have* to start with that assumption."[49] Physicist Harold Morowitz: "Only if we *assume* that life began by deterministic processes on the planet are we fully able to pursue the understanding of life's origins within the constraints of normative science."[50] For what possible reason are scientists *obligated and required* to start with a totally unproven and unsupported assumption while investigating an historical event that took place billions of years ago?

Something has gone terribly wrong here. It is not the job of a scientist to *make* assumptions; it is the job of a scientist to *test* assumptions. Dr. Christian DeDuve makes an almost identical statement, seemingly oblivious to the inherent intellectual corruption contained in such an approach: "It is now generally agreed that if life arose spontaneously by natural processes — a **necessary assumption** if we wish to remain with the realm of science — it must have arisen fairly quickly, more in a matter of millennia or centuries, perhaps even less."[51]

From which Mt. Sinai was issued the proclamation that all inquiry must be based on *assumptions* that keep us within the "realm of science"? Should our conclusions be ignored if reason, logic, and the results of our inquiries point to an answer that is not within the realm of science; particularly in the case of Origin of Life research where some of the most distinguished scientists of the twentieth century have acknowledged that it is a subject beyond the limits of scientific inquiry? Isn't the purpose of all intellectual inquiry to "remain within the realm of *truth*"? If what I am seeking is the truth, what possible reason would I have for caring or "wishing" if it was within the realm of science or not?

Emeritus professor of chemistry at ETH-Zurich, Dr. Pier Luigi Luisi, discussing this very point, implicitly informs us that science has indeed discarded the search for *truth*: "The argument that we *have* to accept the deterministic [and materialistic] view [of the Origin of Life], otherwise we are out of business, may sound naïve and tautological, but actually it is tantamount to our definition of science."[52]

No, Professor Luisi, it does not *sound* "naïve and tautological"; it *is* naïve and tautological! In simple terms, your definition of science means the following: We don't care if a particular proposition is true of not, we just care if it fits into our naïve, tautological, and self-serving definition of science. How refreshingly spot-on is the observation that that for materialistic scientists, the Origin of Life has been downgraded from a sublime discussion about the *truth* to a trivial feud about the definition of science.

It now becomes clear that the agenda and worldview at stake in Origin of Life research is Scientific Naturalism, the atheistic belief that nothing exists other than the material universe and that everything ultimately has a physical/material/scientific explanation. For many scientists this ideology holds the identical place in their psyche, and fulfills the same inner purpose as religious gospel.

We now can understand why Dr. Kauffman and his colleagues refuse to acknowledge that they are *not* investigating *how* life emerged through a naturalistic process, but rather, *if* life emerged through a naturalistic process, because along with diplomats and attorneys, they understand very clearly that whole universes can hang on the placement of a single comma, period, or word. With the addition of the word "if," the game has been irrevocably transformed. The existence of God the Creator and all that potentially implies now becomes a very real consideration and possibility. It is this very possibility that they are determined to avoid, even if it means compromising intellectual integrity and being forced to concoct some very bizarre statements when touching upon the issue of intelligent design or Divine creation. I present below excerpts from the writings of eight distinguished scientists on the Origin of Life to illustrate this point.

Dr. Christian DeDuve — Biochemist and Nobel Laureate (1917–2013)

In his book, *The Genetics of Original Sin*, Dr. Christian DeDuve writes:

> [We have no naturalistic explanation for] the origin of life, which is unknown so far. It thus remains permissible...that life was flipped into being by a Creator... As long as the origin of life can't be explained in natural terms, the hypothesis of an instant Divine creation of life cannot objectively be ruled out. But this hypothesis is sterile, stifling any attempt to investigate the origin of life by scientific means. The only scientifically useful hypothesis is to assume that things, including the origin of life, can be naturally explained.[53]

Let's carefully analyze what this Nobel Prize-winning researcher has written:

- "It thus remains permissible...that life was flipped into being by a Creator... As long as the origin of life can't be explained in natural terms, the hypothesis of an instant Divine creation of life cannot objectively be ruled out." I call the reader's attention to the phrase: "the hypothesis of an instant Divine creation cannot *objectively* be ruled out." "Objectively"? — As opposed to what? Daydreaming? Fantasizing? Wishful thinking? Is there anything besides *objectivity* that should interest a scientist or anyone seeking the truth for that matter? How gracious of Dr. DeDuve to declare it "permissible" to think about the origin of life "objectively"!
- "But this hypothesis is sterile, stifling any attempt to investigate the origin of life by scientific means." Why must the notion of a creator of life be accompanied by a begrudging caveat that it is a "sterile" idea? Could someone please explain to me why an "objectively" reasonable explanation for a particular phenomenon should be labeled as "sterile"? The notion of intelligent alien life in a far-off galaxy generates delirious excitement in the world of science, but why should the real possibility of a Divine creation deserve a less enthusiastic response? How would this "objective" consideration prevent scientists from investigating the plausibility of a naturalistic, unguided process that could lead from non-life to life? Who or what would stop them? Does Dr. DeDuve perhaps fear that once scientists seriously contemplate the possibility of special creation, they will realize it is the obvious answer?

- "The only scientifically useful hypothesis is to assume that things, including the origin of life, can be naturally explained." Is the goal to discover what is "scientifically useful" or to find the truth? Since when does scientific utility trump the search for truth?

Elsewhere, Dr. DeDuve also makes a particularly bewildering statement about the Origin of Life:

> Even if life came from elsewhere, we would still have to account for its first development... How this momentous event happened is still **highly conjectural, though no longer purely speculative**.[54]

I must admit that I was perplexed in trying to understand exactly what Dr. DeDuve meant by "highly *conjectural*, though no longer purely *speculative*," so I decided to research it a little:

- **Speculation***: conjectural consideration of a matter; conjecture or surmise: *a report based on speculation rather than facts*. **Synonym**: conjecture
- **Conjecture***: 1. the formation or expression of an opinion or theory without sufficient evidence for proof. 2. an opinion or theory so formed or expressed; guess; speculation. **Synonym**: speculation

Now I understand. He didn't mean that it's *highly* ***conjectural****, though no longer purely* ***speculative***; he really meant that it's *highly* ***speculative****, though no longer purely* ***conjectural***...or maybe he meant to say *we're shooting in the dark*. All joking aside, it is clear that DeDuve is telling us that "how this momentous event happened" is both highly speculative *and* highly conjectural (in other words, *we're shooting in the dark*). DeDuve wrote this in 1995. In 2006, Richard Dawkins candidly informed us that "the origin of life is a flourishing, if *speculative* subject for research."[55] Interestingly enough, Francis Crick wrote the following in his book *Life Itself*, in 1981: "Every time I write a paper on the origin of life, I determine I will never write another one, because there is too much speculation running after too few facts."[56]

---

* From Dictionary.com

I guess in the twenty-five years between 1981 and 2006 not much was going on (besides a lot of speculating). In fact, the University of Texas at Dallas felt it necessary to state at the end of their online lecture entitled "The Origin of Life on Earth": "Final Caveat: Almost all of this section is highly conjectural. Read it with this in mind."[57] The wry comment on this subject by the late Dr. Michael Crichton is worthwhile noting at this point: "Most areas of intellectual life have discovered the virtues of speculation, and have embraced them wildly. In academia, speculation is usually dignified as theory."[58] In short, the present scenarios hypothetically describing the origin of life are not in the category of science; they are in the category of speculation and conjecture, which is exactly what science is not. Just because the person speculating and conjecturing happens to be a scientist does not magically turn his words and thoughts into science.

Dr. P. Z. Myers, Associate-Professor of Biology at the University of Minnesota

P. Z. Myers is a rather high-profile, outspoken advocate of atheism, and the writer of a popular blog called *Pharyngula*. In a lecture entitled "Design and Chance," given at the Atheist Alliance International in Burbank, California, in 2009, Myers directly confronted the Argument from Design and the origins of the first living cell. He began by giving a brisk summary of Intelligent Design theory: "The core of the argument is this: (a) Complexity can only be created by a designer; (b) Biology is really complex; (c) Biology was created by design."[59] He then posed the following to his audience:

> What about the whole complexity issue? We biologists will freely admit that things are really complicated inside the cell. So how do we explain it? Don't we have to resort to a Creator? And we say, of course not. **There's lots of things that are very complicated [and aren't the result of an intelligent creator]**. I'll show you an example here.[60]

At this point in his PowerPoint presentation, there is a photograph showing a rather large pile of driftwood along what is obviously a coastline. Myers informs us that it is Rialto Beach in upper Washington State. He continues:

> And this is a very common thing along beaches...driftwood. You find these walls of driftwood between you and getting down to the beach, real walls, **very complicated walls. It has been constructed; who did it**? We know the answer: natural processes did it. We don't need a designer to build this kind of wall. **This is complex**, you simply can't deny it. If I turn the projector off would you be able to draw it? No.[61]

In all honesty, the first time I saw the lecture, I was a little shocked at this comparison between the "complexity" of a living bacterium and the "complexity" of a pile of driftwood on a beach. It even crossed my mind that perhaps he was making some kind of joke. To be fair, there is an aspect of truth in what Dr. Myers asserted. From a purely mathematical perspective and mathematical use of the word, the pile of driftwood is "complex" in the sense that there is no simple algorithm that could describe how the pieces of wood are sitting and resting on one another. However, that fact is completely irrelevant to the Argument from Design and certainly to anything else that I've presented here. There isn't anyone (including Dr. Myers) who would not admit that despite its mathematical "complexity," it is immediately recognizable for exactly what it is: a random pile of driftwood. If piles of garbage washed up on the beach they would also be mathematically "complex" and also would be nothing more than random piles of garbage.

On the other hand, the bacterium is *functionally complex*. Its complexity works toward a specific purpose. *Functional complexity* is the difference between a pile of driftwood and a wood cabin built out of driftwood situated on a bluff overlooking the beach. Everyone recognizes immediately that the cabin is the result of conscious, intelligent intervention and likewise everyone recognizes that the pile of driftwood is the result of random, unguided forces. This point seemed so obvious that I was quite puzzled as to where Myers was headed with the presentation. The reader can judge for his or herself, but in my opinion it became even more confused and convoluted.

A photograph of an expertly-constructed brick wall surrounding a garden flashes on the screen. Dr. Myers continues:

> On the other hand we are familiar with this kind of wall. So this is also a wall, it's one that we can recognize that has a **specific purpose**, that was built by human agents, and I'd have to say that of these two walls, **which one is simpler**? The human built one... When we look at natural walls [i.e., driftwood] what we discover is natural things are built by chance and necessity, they are **functionally unspecified**, there's nothing that says that a pile of driftwood is a wall...and they tend to be complex. In this sense, complex often means sloppy, but it's still complex. Artificial walls [i.e., the brick wall] are built with intent, they are **functionally very specific**...and relatively simple.[62]

It's difficult to understand why this was presented at the Atheist Alliance International when everything he says agrees with what I've written in the last two chapters of this book. The brick wall, which is "functionally very specific" and has a "specific purpose," is one that we can "recognize... was built by human [intelligent] agents." On the other hand, "natural things" are "functionally unspecified" and although they are "complex" [mathematically], they are "sloppy" and obviously recognizable as being the product of unguided, natural forces. In short, we all immediately recognize the "functionally very specific" garden wall and molecular machinery of the living cell as being the product of intelligent causation; piles of driftwood, garbage, and chemical sludge are immediately recognizable as being random piles of junk. Why then is Dr. Myers an atheist?

In May 2011, I wrote a column for the Algemeiner.com website offering an analysis and critique of the aforementioned lecture delivered by Dr. Myers at the convention in Burbank. Within twenty-four hours, Dr. Myers had posted a reply to my column on his *Pharyngula* blog entitled: "I am lectured in logic by a man who believes in invisible magic men in the sky." (Just for the record, I don't believe in invisible magic men in the sky.) The gist of Myers' response is that I misrepresented or misunderstood his point: "Nowhere in that talk do I claim that a pile of driftwood is analogous to a cell. I think there's a rather huge difference between a cell and a pile of debris...I was making a different point."[63]

Up until this point in his response we are in agreement. A pile of driftwood is clearly not analogous to a living cell and the cell has nothing in common with piles of debris. The next thing that appears in his post is a reproduction of the famous Nike "swoosh" logo. He goes on to explain that the swoosh is "very, very simple" and is "most definitely designed." So far, so good, we are still in agreement. It is from here on that things get very confusing:

> Is it clearer now? We have complicated things that are not designed, and we have simple things that are designed. We also have complicated things that are designed, and simple things that are not. The message you should take away from these examples is that **complexity and design are independent properties of an object**. One does not imply the other. You cannot determine whether something was designed by looking at whether it is complicated or not.[64]

In the paragraph above, Myers identifies four categories:

1. **"[Mathematically] complicated things that are not designed"**: Examples being piles of driftwood, garbage, and scrap metal. Dr. Myers and I are in complete agreement; these are immediately recognizable as being the products of random, unguided forces.
2. **"Simple things that are designed"**: Examples being the Nike swoosh logo and many other corporate logos. A smiley face is another example of a simply designed artifact or symbol. Usually, despite their simplicity, they are clearly recognizable as being the product of intelligent design. There are of course things that are *so simple* that it is unclear whether they are the result of design or random forces; an example would be if you are walking in the forest and you find two twigs that are crossed in the shape of an "x." Perhaps they were placed deliberately and perhaps not. Dr. Myers and I are in complete agreement here also.

3. **"[Functionally] complex things that are designed"**: Examples being a log cabin, brick wall, cell phone, automobile, etc. Again, we are in complete agreement; all of these are obviously the result of intelligent creative activity.
4. **"Simple things that are not [designed]"**: Examples being twigs crossed like an "x," a cloud that sort of looks like an elephant, patterns of lines in a sand dune, patterns in snowflakes — complete agreement, all of these are clearly the result of unguided forces.

Where then do we disagree? Amazingly enough, the only category on which we disagree is the one category that Dr. Myers does not mention: functionally complex objects — like the brick wall, automobile, watch, etc. — that are **not** designed. Dr. Myers addresses this category separately: "Also familiar, I'm afraid, is the usual indignant waffling [by advocates of Intelligent Design] about it being **specified complexity**... I have never seen it [i.e., specified complexity] operationally defined.[65]

I am at a complete loss as to why the concept of "specified" or "functional" complexity is so puzzling for Myers. It's the difference between the driftwood and the cabin; it's the difference between a pile of scrap metal and a battle-tank; it's the difference between a batch of sludgy chemical goo and a living bacterium. What is so difficult about that to understand? What is even more perplexing is that in the original Atheist Alliance International lecture, *Myers himself* explains the obvious difference between a structure that is "functionally unspecified" and one which is "functionally very specific." At twelve-and-a-half minutes into his lecture, the picture of the aforementioned expertly built brick wall appears on the screen of his Power Point presentation. I repeat verbatim Myers' comments:

> On the other hand we are familiar with this kind of wall. So this is also a wall, it's one that we can recognize that has a **specific purpose** that was built by human agents... Artificial walls [i.e., the brick wall] are built with intent; they are **functionally very specific**...and relatively simple.[66]

It's clear that Dr. Myers understands quite well the meaning of "functional complexity," "functionally very specific," "specified complexity," and "specified purpose." I find it distasteful to have to lecture Dr. Myers in logic, but his response is as logically incoherent as his original presentation. In order to effectively refute my thesis, he would need to give examples of *functionally very specific* or *functionally complex* structures like the brick wall, cabin, or watch that are the result of random, unguided processes. He fails to provide even one such example and for a very good reason: *there are no such examples.*

Myers did not earn a PhD in biology and become an associate professor by being unintelligent. The only explanation for such a profoundly flawed and muddled attempt at refuting the Argument from Design is that Myers is operating under the same faith system as the scientists above; he is psychologically incapable of considering the existence of a Creator of life.

Dr. Robert Shapiro (1935–2011) — Professor Emeritus of Chemistry at New York University

From Shapiro's book, *Origins: a Skeptics Guide to the Origin of Life on Earth*:

> One favorite analogy [of believers] involves the discovery of a watch... It would function only if its components had been put together...by a watchmaker... Similarly, the existence of bacteria and other living beings, all of which are much more complex than a watch, implies the existence of a creator... **We will not take this escape route in our book, for we are committed to seeking an answer within the realm of science**... A being with the capacity to create a watchmaker would be the most complex of the lot. **By following this line of reasoning, we have made our problem more difficult rather than simpler, and we can resolve it only by introducing supernatural forces. We must look for another solution if we wish to remain within science.**[67]

In this passage, Shapiro brazenly throws the quest for truth under the bus. Why does he describe the notion that life was created as an "escape route"? It is because he is "committed" to finding a scientific answer and is prepared to make whatever sacrifices are necessary to "remain within science." Isn't it just as foolish and mistaken for a scientist to believe that his "commitment" to finding a scientific answer magically creates a scientific reality, as it is for a theologian to believe that his "commitment" to finding a religious answer magically creates a spiritual/metaphysical reality? The only "commitment" worth making is to finding the truth. In fact, in order to "remain within science," Shapiro is prepared to ignore what he describes in his own words as a perfectly good line of reasoning; in other words, a valid line of reasoning that leads one to an answer that is outside of science is forbidden. Intellectual integrity has been abandoned to preserve a commitment to the agenda of Scientific Naturalism.

The following email from Shapiro to Dr. "Skip" Evans of the NCSE appeared on the Panda's Thumb website:

> Dear Mr. Evans,
> ...I agree with him [Professor Michael Behe, a prominent proponent of Intelligent Design theory] that conventional origin-of-life theory is deeply flawed. I disagreed with him about the idea that one needed to invoke an intelligent designer or a supernatural cause to find an answer. I do not support intelligent design theories. **I believe that better science will provide the needed answers.**[68]

Is Shapiro's rejection of Intelligent Design based on rigorously obtained experimental evidence? Compelling logic and reason? Conclusive evidence from some other source? It is none of the above. Shapiro "believes" — how scientific! — that science will provide the answers.

In the following passages from Dr. Shapiro's book, *Planetary Dreams*, the parts upon which I will comment are italicized:

> *The discovery of a second separate origin of life within a single Solar System would strongly suggest that the laws of nature include some principle (I shall call it the Life Principle) which favors the generation of life*...The entire history of the Universe can then be interpreted in terms of a process called Cosmic Evolution. A new vision can be built around this concept that provides a sense of purpose for our own future.

> *The weakest point in this belief structure is our lack of understanding of the origin of life.* No evidence remains that we know of to explain the steps that started life here...*If a broadly based Life Principle exists...then some signs of its handiwork should be detectable on the other worlds of our system...*[69]

Is There Such Thing as a "Life Principle"?

"The discovery of a second separate origin of life within a single Solar System would strongly suggest that the laws of nature include some principle (I shall call it the Life Principle) which favors the generation of life."

Let's review the simple facts: No scientist has any real idea how life started on the planet Earth. Dr. Stuart Kauffman went so far as to label one who claims to know "a fool or a knave." If life were found on another planet, *it would not change anything at all*. Scientists would still be completely baffled as to how life started; the only difference would be that now life on *two* planets would be inexplicable. Until scientists have demonstrated empirically that the laws of physics and chemistry can account for the natural formation of a fully functioning bacterium, the only reasonable conclusion would be that the Creator also created life on a second planet. Or alternatively, it would be at least as reasonable a possibility as the one presented by Shapiro.

I have another question for Dr. Shapiro: Why would you propose an unproven, non-scientific, mystical theory like "The Life Principle" and present it as if it were science? If you want to present your own personal stream of consciousness, please go ahead, but why misuse your impressive credentials in chemistry by trying to palm this off as scientific theory? One could suggest this is an example of what Niles Eldredge was referring to when he stated that some scientists think they have "special access to the truth...throwing down scientific thunderbolts from Olympian heights."

"The weakest point in this belief structure is our lack of understanding of the origin of life...if a broadly based life principle exists...then some signs of its handiwork should be detectable on the other worlds of our system."

Dr. Shapiro states that the "*weakest* point in this belief structure is our lack of understanding of the origin of life...etc." Let's briefly review the actual four points that are the *content* of this "belief structure" and then examine the "weak points":

1. The discovery of life on another planet. *Weak point*: There is no compelling evidence at all that life exists on other planets, and even if there were, there would still be no scientific clarity as to how it got there.
2. The existence of something called "The Life Principle" that favors generation of life. *Weak point*: There is no evidence of something called "The Life Principle" that drives inanimate matter to organize itself into living organisms.
3. An understanding of how this "Life Principle" causes life to arise naturally. *Weak Point*: There is no evidence that life can arise naturally from inorganic matter.
4. Based on (1), (2), and (3), understanding the history of the universe and societal development according to a process called *Cosmic Evolution* and a "new vision" built around this Cosmic Evolution that "provides a sense of purpose for our own future." *Weak point*: Since the concept of "Cosmic Evolution" is built on the truth of (1), (2), and (3), there is no evidence that such a thing as "Cosmic Evolution" exists.

Besides that, Shapiro's "belief structure" is pretty sound.

At the beginning of 2006, Dr. Shapiro predicted the following:

> We shall understand the Origin of Life within the next five years... two very different groups will find this development dangerous... many scientists have been attracted by the RNA World theory*... they would not be pleased [if Shapiro's "Metabolism First" theory of Origin of Life turned out to be true]... Those who advocate creationism and intelligent design would feel that another pillar of their belief system was under attack... A successful scientific theory would leave one less task for God to accomplish: the Origin

* The RNA World Theory and Metabolism First Theory are two of the popular speculative theories to explain the Origin of Life. There are a number of others, including Deep Sea Thermal Vents, Primordial Soup, and Life from Outer Space.

> of Life would be a natural (and perhaps frequent) result of the physical laws that govern this universe...[70]

I have to admit that it was quite bold of Dr. Shapiro to predict a naturalistic explanation of the Origin of Life by the year 2011, and even to predict who would be upset at this discovery. On January 22, 2008, two years after this prediction, Dr. Leslie Orgel published an article claiming that Shapiro's Metabolism First theory was based on "if pigs could fly" chemistry.[71] To complicate matters further, an article on the *Science Daily* website (Jan 9, 2010) appeared with the following headline: "New Study Contradicts the 'Metabolism First' Hypothesis." My question is: was anyone upset that the Origin of Life was *not* understood by January 2011? It didn't seem to make any difference at all. No committed believer in the dogmas of Scientific Naturalism would allow something that trivial to shake their faith.

Dr. Paul Davies of Arizona State University Loyally Repeats the Catechism

> In the coming chapters I shall argue that it is not enough to know how life's immense structural complexity arose; we must also account for the origin of biological information. As we shall see, scientists are still very far from solving this fundamental conceptual puzzle... **However it is the job of scientists to solve mysteries without recourse to Divine intervention**.[72]

Note: It is not the job of scientists to seek the *truth*, it is their job to "avoid recourse to Divine intervention," and defend the faith of Scientific Naturalism to the bitter end.

Dr. Jerry Coyne, Professor of Evolutionary Biology at the University of Chicago

In a post on his widely read blog, *Why Evolution Is True*, Dr. Coyne had this response to the Argument from Design as it relates to the mystery of the Origin of Life:

> Nope, we don't yet understand how life originated on Earth, but

> we have good leads, and abiogenesis [life from non-life] is a thriving field. And we may never understand how life originated on Earth, because the traces of early life have vanished. We know it happened at least once (and that all species descend from only one origin), but not how. I'm pretty confident that within, say, fifty years we'll be able to create life in a laboratory under the conditions of primitive Earth, but that, too, won't tell us exactly how it did happen — only that it could. And if it could, then we needn't [*sic*] postulate a much less parsimonious celestial deity, especially one who forbids you to eat bacon, or enjoy meat and cheese at the same meal.[73]

"Nope, we don't understand how life originated on Earth, but we have good leads..." Perhaps Dr. Coyne should share some of these "good leads" with:

- Origin of Life expert Dr. Paul Davies, who candidly informed us about the origin of life: "How? We haven't a clue."
- Dr. George Whitesides of Harvard University, one of the world's greatest living chemists, who made the almost identical statement: "How? I have no idea."
- Dr. Andrew Knoll of Harvard University: "We don't know how life started on this planet. We don't know exactly when it started, we don't know under what circumstances...I don't know if [we will ever solve the problem]...I imagine my grandchildren will still be sitting around saying it's a great mystery."
- Dr. Freeman Dyson: "There is an enormous gap between the simplest living cell and the most complicated naturally occurring mixture of nonliving chemicals. We have no idea when and how and where this gap was crossed."
- Dr. Milton Wainwright: "Are we getting any closer to an understanding of the origin of life?... It is not merely that biology is scratching the surface of this enigma; the reality is that we have yet to *see* the surface!"
- Dr. Eugene Koonin: "The origin of life field is a failure."
- All the other scientists cited in Chapter 3.

"*And we may* ***never*** *understand how life originated on Earth, because the traces of early life have vanished. We know it happened at least once...but not how.*" At least there is something that Dr. Coyne and I agree on: We definitely do know that life originated *at least once*. I can't say it's an insight that takes my breath away, but I do agree. Simple question for Dr. Coyne: If you admit that you don't know how life originated and that you may never understand how life originated, how do you know that the "traces of early life have vanished"? Actually that's easy, because no one has ever discovered the slightest evidence of early life. Let's rephrase the question: If there is no evidence of early life, how do you know there ever was such a thing as "early life"?

"*I'm pretty confident that within, say,* ***fifty years*** *we'll be able to create life in a laboratory under the conditions of primitive Earth, but that, too, won't tell us exactly how it did happen — only that it could.*" We've already seen that the lexicon of Origin of Life research includes such "scientific" terms as *belief*, *surmise*, *speculate*, *conjecture*, *philosophical presuppositions*, *miraculous*, "*Monte Carlo game*," and "*article of faith*." Dr. Coyne has now added: "I'm pretty confident." There is a Talmudic dictum: "One who wishes to lie always makes sure his witnesses and evidence are far away." Dr. Coyne offers us nothing in the way of solid reasoning or scientific evidence, but is very careful to make sure that his witnesses and evidence are very far away, actually *fifty years away*, when he claims everything will become clear. The impotence of such an argument speaks for itself.

What's even more interesting is that in another post on his blog **two years later**,[74] Dr. Coyne wrote the following: "How the unique properties of life originated from inert matter is still one of the great unsolved problems of biology... Perhaps we'll never know precisely how life began, for it happened in the distant past and involved chemical reactions that could not fossilize." Again, if it is such a great mystery and happened in the distant past, how does Dr. Coyne *know* that it involved chemical reactions that could not fossilize? He continues: "But I have *confidence* [that] life originated naturally and not through God's fiat [and] that we will show this was possible within fifty years or so by demonstrating the evolution of life-like systems in the laboratory under primitive earth conditions." While it's gratifying to see that his confidence hasn't waned, it is anything but gratifying to note that two years later, instead of being at forty-eight years and counting, he is stuck in a time warp and is still writing about discovering a plausible scenario for a naturalistic origin of life within **fifty years**! With this approach he could go on *ad infinitum* or *ad nauseam*, take your pick.

Imagine a district attorney going before a judge with a motion to deny bail because the defendant is a very dangerous person. When asked for evidence to back up his claim he replies, "All the evidence has vanished, but I'm pretty confident that within fifty years we'll have all the evidence we need." The judge would likely throw him in jail for contempt of court. Jerry Coyne is in contempt of scientific integrity, intellectual integrity and the search for truth.

Dr. Ken Nealson, who was co-chairman of the Committee on the Origin and Evolution of Life for the National Academy of Sciences once stated: "Nobody understands the origin of life, if they say they do they are probably trying to fool you."[75] If you Google the following phrase, "People who are trying to fool you about the Origin of Life," you will see a picture of Dr. Jerry Coyne.

Dr. Addy Pross, Professor of Chemistry, Ben-Gurion University, Israel

Doctor Pross, in his book, *What is Life?: How Chemistry Becomes Biology*, not only confronts the Origin of Life dilemma head on, but even seems to present a compelling case for the existence of a Creator. He hypothetically proposes the discovery of a fully functioning refrigerator — with a solar panel for electricity — in the middle of a large field. There is even cold beer inside:

> But the mystery of how it got there in the middle of the field remains. Who put it there? And why? **Now if I told you that no one put the refrigerator there, that it came about spontaneously through natural forces, you would react with total disbelief. How absurd! Impossible!** Nature just doesn't operate like that! Nature doesn't spontaneously make highly organized....purpose-

> ful entities... Nature...pushes systems...toward disorder and chaos, not toward order and function.
> **The simple truth is that the most basic living system, a bacterial cell, is a highly organized...functional system...which mimics the operation of the refrigerator, but is orders of magnitude more complex!...** [A]bacterial cell involves the interactions of thousands of different molecules and molecular aggregates...every living cell is effectively a highly organized factory...This miniature factory converts [raw materials] into functional components.
> Paradoxically despite the profound advances in molecular biology in the past half century, we still do not understand what life is, how it relates to the inanimate world, and how it emerged.
> And here precisely lies the [origin of] life problem... It is not just common sense that tells us that highly organized entities don't just spontaneously come about. Certain basic laws of physics preach the same sermon — systems tend towards chaos and disorder, not toward order and function... Biology [i.e., a naturalistic origin of life] and physics seem contradictory, quite incompatible. **No wonder the proponents of Intelligent Design manage to peddle their wares with such success!**(*x–xii*)

Dr. Pross and I are in agreement on just about everything he has written:

- The notion of a naturalistic origin of life seems "absurd" and "impossible."
- Natural forces push in the direction of chaos and disorder not towards super-sophisticated functionally complex systems like a refrigerator and certainly not a bacterium, which is many orders of magnitude more complex.
- Not only is the notion of a Creator of life the obvious "common sense" conclusion but it is also in consonance with basic principles of physics and mathematical probability.
- The notion of a naturalistic origin of life seems to *contradict* the most fundamental principles of physics.

If Intelligent Design is the common sense answer and in line with physics and mathematical probability, why does Dr. Pross feel it necessary to pejoratively describe its proponents as "peddling their wares"? In truth, of course, it is just the opposite. It is the anti-common sense, anti-principles-of-physics, anti-mathematical-probability atheistic scientists who are "peddling their wares." It is terribly disappointing when a man of Doctor Pross' caliber makes such a blatantly self-contradictory and non-scientific statement in order to pay homage to the faith of Scientific Naturalism.

Dr. Leslie Orgel (1927–2007), Senior Fellow and Research Professor, Salk Institute — San Diego, California

In an article entitled, "The Origin of Life on Earth," Dr. Orgel, a long-time collaborator of Francis Crick, tells us:

> When the earth formed some 4.6 billion years ago, it was a lifeless, inhospitable place. A billion years later it was teeming with organisms resembling blue-green algae. How did they get there? How, in short, did life begin?... Before the mid-seventeenth century, most people believed that God had created humankind and other... organisms... Darwin, bending somewhat to the religious biases of his time, posited in the final paragraph of *The Origin of the Species* that the "Creator" originally breathed life "into a few forms or into one." Then evolution took over...
>
> In private correspondence, he suggested life could have arisen through chemistry, "in some warm little pond, with all sorts of ammonia and phosphoric salts." For much of the twentieth century, origin of life research has aimed to flesh out Darwin's private hypothesis — to elucidate how, **without supernatural intervention**, spontaneous interaction of the relatively simple molecules dissolved in the lakes or oceans of the prebiotic world could have yielded life's last common ancestor.[76]

Please note that Orgel does not say that Origin of Life research has aimed to discover if Darwin's private hypothesis has any validity, or if it can empirically be shown to be true. He makes it clear that Origin of Life researchers have declared their allegiance of faith to this proposition and are simply trying to figure out how it happened. What I also find disturbing is that Orgel is less than candid with his readers. He fails to mention, that although "for much of the twentieth century" (and at this point well into the twenty-first century) Origin of Life research has "aimed to flesh out Darwin's private hypothesis," that particular endeavor has until now met with complete failure.

Dr. Orgel, however, has added another dimension to the faith of Scientific Naturalism by invoking the name of Darwin. I guess by tracing the theory of life emerging from non-life back to Darwin, it somehow gives it more...holiness? Amazingly enough, at one time in his career Orgel was prepared to accept the possibility of "men from Mars" as the answer to our question. If you will recall, he co-authored the paper on Directed Panspermia with Francis Crick. It's hard for me to understand why a rational person would consider aliens from space but not bother to consider a Creator.

Orgel was a true believer in Darwin's "private hypothesis" for at least forty years (any allusion to Biblical events purely coincidental). As of yet, the Darwinian Origin of Life messiah has failed to arrive. How many more years are the true believers prepared to wait before *considering* another answer?

Dr. Stephen J. Gould, Renowned Paleontologist, Declares His Faith:

> The earth is 4.6 billion years old, but the oldest rocks date to about 3.9 billion years because the earth's surface became molten early in its history... The oldest rocks...to retain cellular fossils [date] to 3.55 billion years... Thus, life on the earth evolved quickly and is as old as it could be.
>
> **This fact alone seems to indicate inevitability, or at least predictability, for life's origin from the original chemical constituents of atmosphere and ocean**.[77]

How do we know that life originated from the original chemical constituents of atmosphere and ocean in a purely naturalistic process? Because, as Gould tautologically informs us, life is as old as it could be, therefore it obviously originated *inevitably* or *predictably* from the atmosphere and ocean... Is that a fact? Why did Gould overlook the simple reality that when he wrote his article (and it still holds true today), scientists had nothing more substantive than highly speculative ideas about how life could have originated "naturally," that Origin of Life researchers are "baffled" and "perplexed" as to how life started? Maybe the reason that life is "as old as it could be" is because it was created. Maybe life is not "inevitable" or "predictable" at all. After all, nobody was there to see what happened and there is no scientist alive today who would claim, even in his wildest imagination, that he could predict how life would *inevitably* arise under *any* particular set of naturalistic circumstances. However, Gould is forbidden by his faith in Scientific Naturalism from considering such a possibility. Dr. Gould may very well have been a great scientist, but it is clear that his dearly held preconceived notions grossly distorted his objectivity on this issue. The one consolation is that he has lots of other great scientists as company.

Part 2 of Objection #1: Scientists Are Busy Working on the Problem

The Origin of Life Prize is real and the information appears on the www.us.net/life website:

> ***"The Origin-of-Life Prize"***
> **Description and Purpose of the Prize**
> **Lifeorigin.info**
> *"The Origin-of-Life Prize" will be awarded for proposing a highly plausible mechanism for the spontaneous rise of genetic instructions in nature sufficient to give rise to life. To win, the explanation must be consistent with empirical biochemical, kinetic, and thermodynamic concepts...and be published in a well-respected, peer-reviewed science journal. The one-time prize will be paid to the winner(s) as a twenty-year annuity...the annuity consists of $50,000 per year for twenty consecutive years, totaling one million dollars in payments.*

In the arena where the search for truth is played out, scientific or academic credentials *per se* have little or no value. The potent weapons in this battle are not impressive credentials but rather coherent, compelling arguments based on logic, reason, and evidence. Scientists — or theologians and clergyman for that matter — are not granted special pleadings or privileges no matter how many PhDs they have after their name. "We are working on the problem" is a strategy to avoid squarely facing up to (a) the powerful questions and implications raised by the Origin of Life issue, (b) the conspicuous lack of evidence for a naturalistic origin of life, and (c) that there is no scientific light visible at the end of the tunnel (or anywhere else for that matter), and (d) the utter failure of Origin of Life research over the past 150 years.

The fact that until now scientists have failed to present a "highly plausible" explanation for a naturalistic Origin of Life is the atheist's and agnostic's problem, not mine. If you discover a way to empirically and plausibly demonstrate that life could have originated from inorganic matter naturally, please call me immediately (and enjoy the fifty grand a year). In the meantime, I hope you'll excuse me if I don't sit breathlessly by the phone like some teenage girl waiting for someone to call and ask her for a date. In short, the declaration by atheistic scientists that "it is not science" is not a valid *reason* to reject a Creator and Intelligent Design theory; it is nothing more than a hopelessly lame *excuse*.

End Notes

1 http://www.ucsusa.org/scientific_integrity/what_you_can_do/why-intelligent-design-is-not.html#.VKtaWiuUcio

2 http://undsci.berkeley.edu/article/0_0_0/id_checklist

3 From "Letter Concerning 'Intelligent Design' and Education to President George W. Bush," (PDF) (Press release), American Astronomical Society, August 05, 2005, http://en.wikipedia.org/wiki/List_of_scientific_bodies_explicitly_rejecting_Intelligent_design

4 http://en.wikipedia.org/wiki/List_of_scientific_bodies_explicitly_rejecting_Intelligent_design

5 "ASBMB President Writes to President Bush on 'Intelligent Design'" (Press release), American Society for Biochemistry and Molecular Biology, August 04, 2005, http://en.wikipedia.org/wiki/List_of_scientific_bodies_explicitly_rejecting_Intelligent_design

6 *Science and Creationism: A View from the National Academy of Sciences* (Washington, DC: National Academies Press, 2nd Edition), p. 25.

7 Niles Eldredge, *The Monkey Business: A Scientist Looks at Creationism* (Washington Square Press, 1982), p. 27.

8 Niles Eldredge, *Timeframes* (New York, NY: Simon and Schuster, 1985), pp. 46–47.

9 S. J. Gould, "In the Mind of the Beholder," *Natural History*, Vol. 103 (February, 1994), p. 14.
10 "Rockefeller U. Biologist Wins Nobel Prize for Protein Cell Research," *New York Times*, October 12, 1999, p. A29.
11 Martin Enserink, "Peer Review and Quality, a Dubious Connection?" *Science*, Vol. 293 (2187–2188), September 21, 2001.
12 "Nobel Winner Declares Boycott of Top Science Journals," *The Guardian*, December 9, 2013.
13 Frank Tipler,"Refereed Journals: Do They Insure Quality or Enforce Orthodoxy?" *International Society for Complexity, Information, and Design Archives* (June 30, 2003), http://antigreen.blogspot.co.il/2013/03/refereed-journals-do-they-insure.html
14 Ibid.
15 Dr. Juan Miguel Campanario, "Rejecting and Resisting Nobel Class Discoveries: Accounts by Nobel Laureates," *Scientometrics*, November 2009.
16 Ibid.
17 Dr. Bernard Nathanson, *The Hand of God: A Journey from Death to Life* (Washington, DC: Regenery Publishing, 1996), Chapter 7.
18 Dr. Michael Crichton, "Ritual Abuse, Hot Air, and Missed Opportunities," *Science*, March 5, 1999.
19 Yoram Bogacz, *Genesis and Genes* (New York: Feldheim), p. 14.
20 Simon Singh, *Big Bang* (Harper Perennial, 2004), p. 267.
21 Bogacz, *Genesis and Genes*, p. 17.
22 Ibid., p. 23.
23 Reuters, October 5, 2011, http://www.reuters.com/article/2011/10/05/nobel-chemistry-idUSL5E7L51U620111005
24 Bogacz, *Genesis and Genes*, pp. 133–137.
25 Max Planck, *Scientific Biography and Other Papers* (New York: Philosophical Library, 1949), p. 33.
26 Dr. Robert Hazen, *Genesis*, *xiii*.
27 Ibid.
28 Ibid., *xiii, xiv*.
29 Ibid., p. 241.
30 Ibid.
31 Ibid., *xxi, xiv*.
32 Dr. Iris Fry, "Are the Different Hypotheses on the Emergence of Life as Different as They Seem?" *Biology and Philosophy* 10 (Netherlands: Kluwer Academic Publishers, 1995), pp. 389–417.
33 http://en.wikipedia.org/wiki/Falsifiability
34 Fry, "Are the Different Hypotheses..."
35 *Christian Science Monitor*, January 4, 1962.
36 Fry, "Are the Different Hypotheses..."
37 Crick, *Life Itself*, pp. 51–52.
38 Fry, "Are the Different Hypotheses..."
39 Stephen Meyer, *Signature in the Cell: DNA and the Evidence for Intelligent Design* (New York: Harper Collins), p. 244.
40 Crick, *Life Itself*, p. 52.
41 Stephen Shankland and Melanie Austria Farmer, "IBM to research proteins using supercomputer," CNET News, (December 7, 1999), http://www.zdnetasia.com/news/hardware/0,39042972,13022659,00.htm
42 Smith, *Seven Clues*, p. 30.

43 Jacque Monod, *Chance and Necessity: an Essay on the Natural Philosophy of Modern Biology*, 1971 (London: Penguin, 1997, reprint), p. 143.
44 Fry, "Are the Different Hypotheses..."
45 Ibid.
46 Pier Luigi Luisi, *The Emergence of Life: From Chemical Origins to Synthetic Biology* (Cambridge University Press, 2006), p. 9.
47 Kauffman, *At Home in the Universe*, p. 31.
48 Dr. Paul Davies and Phillip Adams, "In Search of Eden," http://www.abc.net.au/science/more-bigquestions/stories/s540242.htm
49 Harold Morowitz, *Beginnings of Cellular Life* (Yale University Press, 1992), p. 3.
50 Dr. Christian DeDuve, "The Beginnings of Life on Earth," *American Scientist* (Sep/Oct 1995), http://www.americanscientist.org/issues/feature/the-beginnings-of-life-on-earth/1
51 Luisi, *The Emergence of Life*, p. 6.
52 Dr. Christian DeDuve, *Genetics of Original Sin* (Yale University Press, 2010), p. 16.
53 DeDuve, "The Beginnings of Life on Earth."
54 Dawkins, *The God Delusion*, p. 164.
55 Crick, *Life Itself*, p. 153.
56 Lecture 18 — The Origin of Life on Earth, http://www.utdallas.edu/~cirillo/nats/day18.htm
57 Dr. Michael Crichton, "Why Speculate," lecture (April 26, 2002), http://www.michaelcrichton.net/speech-whyspeculate.html; http://bevets.com/equotesc.htm
58 P. Z. Myers, "Design vs. Chance," Atheist Alliance International Lecture (2009), YouTube.com
59 Ibid.
60 Ibid.
61 Ibid.
62 P. Z. Myers, "I am lectured in logic by a man who believes in invisible magic men in the sky," *Pharyngula Blog*, May 29, 2011.
63 Ibid.
64 Ibid.
65 P. Z. Myers, "Design vs. Chance."
66 Shapiro, *Origins*, p. 119.
67 www.pandasthumb.org/archives/2005/10/robert-shapiro.html
68 Shapiro, *Planetary Dreams*, p. 26.
69 www.edge.org/q2006/q06_9.html
70 "The Implausibility of Metabolic Cycles on the Prebiotic Earth," Leslie Orgel, published posthumously, Jan 22, 2008, PlosBiology.org
71 Davies, *The Fifth Miracle*, p. 58.
72 Dr. Jerry Coyne, "A Rabbi Proves God," *Why Evolution Is True Blog*, March 7, 2011, whyevolutionistrue.wordpress.com
73 Dr. Jerry Coyne, "Paul Davies, Chemistry, and the Origin of Life," *Why Evolution is True Blog*, January 14, 2013.
74 Robert Roy Britt, "The Search for the Scum of the Universe," May 21, 2002, Space.com
75 Dr. Leslie Orgel, "The Origin of Life on Earth," *Scientific American*, (Oct. 1994).
76 Stephen J. Gould, "The Evolution of Life on Earth," *Scientific American*, October, 1994.

Chapter 5

The Watchmaker Has 20/20 Vision: The Philosophical Objections

Non-theists who have achieved any meaningful level of sophistication are quite aware that scientists have no idea how life began. As our awareness of the complexity of life expands, the bafflement of Origin of Life researchers expands proportionally if not exponentially. That is why for the most part the counter arguments put forth by atheist polemicists who directly address this issue are not based on science. After all, there isn't much to say from a scientific perspective except that "we're clueless but we're still trying (and hoping)." Rather they launch *philosophical* attacks to undermine and deflect the potency of the Argument from Design as it relates to the Origin of Life. This very fact represents a paradigm shift that seems to go unnoticed by most people.

Ideologically committed atheists often pride themselves on their connection to science and the truth they feel it represents. It is quite common for them to frame the battle between believer and non-believer as being between modern, rational *scientific* thinking and medieval superstitious nonsense. The acclaimed work by iconoclastic author and scholar

Dr. David Berlinski, *The Devil's Delusion: Atheism and its Scientific Pretensions*, is clearly a commentary on this attitude. However, the primacy of these philosophical counter arguments in atheistic thinking is an implicit admission that, when all is said and done, the battle over the existence of God the Creator has nothing to do with science.

The irony that escapes many atheists is that once we enter the *philosophical* arena, science and scientific credentials carry no authority, confer no intellectual edge, and are for the most part, irrelevant — this applies even to a Nobel Prize-winning level of scientific expertise. The fact that you can throw a ninety-five mph fastball doesn't matter on the football field, and your lab coat, beakers, pipettes, and PhD in microbiology don't matter in a clash of philosophical ideas and arguments. This is not a battle between those who support science and those who are anti-science; it is a battle between theistic *philosophers* and atheistic *philosophers*. Seeing as this is the case, let the games begin.

The Argument from Ignorance/God of the Gaps

The most common philosophical objection of the non-believer is the accusation that what I've presented to you in the last two chapters is a type of flawed logic called an *Argument from Ignorance* (AFI). For those who are unclear, an AFI is the following:

> Imagine I am searching for an explanation of Phenomena X. After a thorough investigation I am (a) either unable to come up with an explanation or (b) I eliminate Explanation A as a possibility. If I were to then go ahead and say: Because I have not been able to discover an explanation or because it is not Explanation A, I therefore conclude that it must be Explanation B; that would certainly be an Argument from Ignorance. The fact that I don't know or that it is not Explanation A is not evidence that it is Explanation B. For all I know it could be Explanation C, D, E, F, etc. If I want to assert that Explanation B is the answer, I need positive evidence that it is true, not simply ignorance of an alternative answer.

As applied to our case, critics claim I am unjustified in concluding that life was the result of Intelligent Design because my only supporting evidence for its truth is the *scientific ignorance* regarding a naturalistic process that could bridge the gap between non-life and life. In other words, just because *we don't know*, Intelligent Design is not a legitimate conclusion. Dr. Jerry Coyne put it this way: "Here [is the theist's argument] in one sentence: Because we don't understand how life originated on Earth, God must have done it,"[1] hence, an Argument from Ignorance. This same objection in a slightly different form is commonly called "The God of the Gaps Argument":

> Rabbi, just because science has not yet found an explanation does not mean that an intelligent creator or God did it. There is simply a gap in scientific knowledge and understanding and you have used that gap as an excuse to jam God or a Creator into the picture.

Hence, God of the Gaps. Both of these objections are baseless, as we will see.

SETI

There exists a long-running scientific project called SETI — the **S**earch for **E**xtra-**T**errestrial **I**ntelligence. Scientists scan the sky with radio-telescopes hoping to detect patterns of radio waves that would indicate an intelligent source. Right from the start, then, it becomes obvious that scientists, in fact all people, have *actual criteria* for determining whether or not a phenomenon is the result of intelligent causation. If not, the entire project would be an exercise in futility. Let us try to articulate what these criteria are and see if a living bacterium fits the bill. Until the present time the project has failed to discover anything significant. Imagine, though, that these SETI scientists detected the following:

1. Four distinct patterns of signals, each lasting approximately three seconds,
2. These four signals were repeated over and over again a total of 11,000 times in no particular order,
3. It was clear beyond any reasonable doubt that the source of these signals was a galaxy a million light years away, therefore the source, whatever it was, could not possibly be human.

At first, no one was quite sure about the significance of these signals and it was decided to process them through a computer to see if they matched any known intelligible patterns. It was discovered that if the letters A, T, G, and C were assigned to each of the four patterns, the signals were a perfect match for 11,000 nucleobase-pairs on the DNA of a bacterium called a *mycoplasma genitalium*, which like all DNA consists of **A**-Adenine, **T**-Thymine, **G**-Guanine, and **C**-Cytosine. These particular 11,000 base-pairs coded for the synthesis of a molecular machine called an RNA polymerase, which plays a crucial role in copying information from the unwound strands of DNA so that the cell can produce the proteins it needs to survive. Would that not be undeniable evidence of intelligent causation?

While I am certain that intuitively every reader would agree with the validity of this conclusion, it is important that we state explicitly why this conclusion is inescapable. It is because these radio signals convey and represent significant, coherent, intelligible coded information. That fact is the entire difference between random, meaningless, and insignificant radio signals — which is what SETI scientists have encountered until the present time — and these radio signals that indicate *intelligence*. Imagine then the following conversation taking place between two SETI scientists:

Scientist #1: How do you know that the source is an intelligent alien life form? Maybe there is some naturalistic, unguided process completely within the laws of physics that is the cause of these radio transmissions?

Scientist #2: (*Incredulously*) What unguided physical process do you know of that could produce radio signals corresponding to series of nucleobases on DNA to produce an RNA polymerase?!

Scientist #1: Gotcha! An argument from ignorance! Just because *we don't know* of any such process, just because there is *scientific ignorance*, does that mean there must be an intelligent, creative force behind these signals? After all, did you meet these aliens? Do you know who, where, or what they are? Do you have any independent evidence that they really exist? As Dr. Jerry Coyne has put it, "Here is the alien-believers argument in one sentence: Just because scientists don't know how these radio signals originated, aliens must have sent them."

Is the conclusion then that the source of these signals was intelligent alien life an Argument from Ignorance? Of course not!

In fact, we are not ignorant at all regarding the causes and sources of specified coded information. *Any and all* language, both written and spoken, is specified coded information. There is no difference between specified information that is in a language consisting of a four "letter" radio signal that codes for the instructions to produce functional proteins or in a language that consists of a four-letter chemical/molecular alphabet on the double helix of DNA that codes for the instructions to produce the same functional proteins.

In terms of identifying intelligence as the source of information, it makes no difference whether the information is conveyed in Morse code, English, French, sign language, drum beats, or smoke signals. The only source of such phenomena is creative, conscious, intelligent activity. We recognize this immediately. Random radio signals from outer space signify nothing as do random strings of nucleobases. Unguided forces produce gibberish — like the monkey at the word processor — not intelligent, coherent, useful information. This has been confirmed by all human experience, observation, testing, and supported by basic principles of physics and mathematical probability, as we pointed out in the third chapter. In other words, it is a conclusion and argument based not on ignorance, but on a wealth of experience, observation, knowledge, and clarity.

The reason we conclude that the source of these signals is intelligent activity is not *just* because we don't know of any naturalistic process that could produce such signals (which of course there isn't), nor do we conclude that the encyclopedic amount of digital information in DNA is the result of intelligence just because we *don't know* of any unguided process that could digitally encode and process such information (which of course there isn't). It is because we do know *exactly what does* produce these types of specified coded messages and we also know *exactly what does not* produce such coherent information. This *knowledge* is so clear in our minds that we don't even consider another possibility.

The arrangement of the magnetic letters on the refrigerator door *itself* is the undeniable evidence of the presence of intelligent activity, the *digital information and instructions themselves* are the evidence for the presence of intelligence in the encoding of DNA, and the *intelligible information itself* in the radio transmissions is the irrefutable evidence for the existence of the *totally unknown* intelligent cause of the message. We have the ability to recognize, discern, and identify intelligent causation whether it be animal (like a spider's web or beaver dam), human, or totally other. We know nothing about the source of these radio signals other than the fact that whoever sent them is capable of intelligent, creative activity. Our lack of understanding about the source does not prevent us from recognizing the intelligent causation involved.

It is patently obvious that *all* scientists agree that we are able to recognize and identify intelligent causation, whether it be in the form of radio signals from outer space or complex digitally encoded information on a strand of DNA. The taken-for-granted acknowledgement of this fundamental fact is the *foundation of the entire SETI project*. If we do not have the ability to recognize intelligent causation when we see it, what was the point of investing hundreds of thousands of man-hours of work and untold millions of dollars in a project called the Search for Extra-Terrestrial *Intelligence*?!

Aliens of the Gaps?

> Rabbi, you're telling me that just because there is a *gap* in scientific knowledge and understanding and we haven't yet discovered a naturalistic, unguided process completely within the laws of physics and chemistry that could account for these radio signals, you posit the existence of a completely unknown alien civilization of which you have not a shred of independent evidence to confirm its existence?! Do you also believe in Superman from the planet Krypton (*smirk, smirk*)? Rabbi....that's the old Aliens of the Gaps Argument! Could anything be more ridiculous? To claim that Intelligent Design for the Origin of Life is a God of the Gaps Argument is equally ridiculous.

In fact, what the atheist proposes is much worse than an Argument from Ignorance. (When I say "atheist," I mean those who actively deny the existence of God the Creator, as opposed to those who adopt an agnostic-like position.) What the atheist proposes is the following: "Well folks, we admit that we don't have the slightest clue how the enormous, yawning chasm between life and non-life was crossed through a naturalistic, unguided process. Therefore, we somehow conclude that there ***cannot*** be a Creator." Try making sense out of that logic.

Let's be honest: If someone wanted to assert that these radio signals that conveyed sophisticated information were the result of an unguided, naturalistic process — besides the fact that we'd assume they'd lost their mind — the burden of proof would be on them. If you want to assert that the bacterium, along with its astounding nanotechnology, its genetic information processing system, and the enormous amounts of pre-loaded digitally encoded information, are the result of an unguided process, the extraordinarily heavy burden of proof is on you.

Remember, the denial of Intelligent Design is not due to a lack of perception of design in the living world. This perception of design is so overwhelming that Richard Dawkins quite candidly admitted that he "could not imagine being an atheist at any time before 1859, when Darwin's *Origin of the Species* was published."[2] Christopher Hitchens also acknowledged that before Darwin, the position that the observed order in the universe indicated a designer was "a logical and rational one for its time."[3] Jerry Coyne, Lawrence Krauss, Francisco Ayala, Francis Crick, the Wyss Institute, all acknowledge design; they simply claimed that these are *illusions* of design caused by Darwinian Evolution.

We have shown that invoking Darwinian Evolution to deny Intelligent Design is a fatally flawed approach, though not due to a rejection of Darwinian Evolution; on the contrary, for argument's sake we have conceded its truth. It is because the process of Darwinian Evolution is not operative without pre-existing functionally complex molecular machinery, enormous amounts of digitally encoded information and the machinery to store, retrieve, and translate that information. Darwinian Evolution simply begs the question. What we are left with when we contemplate the origin of the first living cell is an obvious and overwhelming perception of design *with no reason to doubt that it is absolutely real*. If you want me to believe that the astounding "transcalculational" sophistication and the levels of functional complexity "that outstrip scientific description" that are apparent in a bacterium and its genetic information system are *illusions of design*, then prove it. Demonstrate to me empirically, logically, or experimentally that they are *illusions of design* and I will believe you. I wish you luck...you will need it.

The Argument from Infinite Possibilities or Atheism of the Gap

I accuse the atheist of routinely using the most egregiously flawed argument of all. Let me begin by citing one of the great intellectuals of the twentieth century, Bertrand Russell, who made the following oft-quoted statement:

> Many orthodox people speak as though it were the business of skeptics to disprove received dogmas rather than of dogmatists to prove them. If I were to suggest that between the Earth and Mars is a china teapot revolving about the sun in an elliptical orbit, nobody would be able to disprove my assertion provided I were careful to add that the teapot is too small to be revealed even by our most powerful telescopes. But if I were to go on to say that since my assertion cannot be disproved, [no one can doubt its truth], I should rightly be thought to be talking nonsense.[4]

Russell is absolutely correct. If I propose some fantastic notion and demand that it be accepted as truth, it is my burden to present the evidence that it is true. The fact that the particular notion cannot be *disproved* is irrelevant. Another way of stating that something cannot be disproved is to say, "well, it's *possible*," or "it's not *impossible*." The fact that it's *possible* or not *impossible* is meaningless. It is obviously *possible* and certainly not *impossible* that not only the teapot but an accompanying set of silver flatware is also in orbit between Earth and Mars. So what?

The notion that the awe-inspiring levels of functional complexity and specified information found in the "simplest" living bacterium is the result of some mysterious, unguided, undirected process is an extraordinary claim. It is so extraordinary that one group of researchers considers it to be a "miraculous" stroke of luck that life came from non-life, and the other camp accepts it, at best as an unproven, non-falsifiable philosophical presupposition, or more accurately, as an article of faith.

Extraordinary claims require extraordinary evidence. As of today, not only is there no extraordinary evidence available, but as we have pointed over and over again, we find just what we would expect: utter bafflement on the part of researchers and no evidence at all that would compel anyone to accept this ridiculous proposition as fact. Atheistic scientists are acutely aware of the difficulties involved in proposing an unguided process able to bridge the gap between non-life and life. However, they seem totally oblivious to the fact that, in keeping with the thrust of Russell's argument, it is their burden to prove it true rather than being my burden to *disprove* the possibility. What we will discover is that the non-theist has latched onto the bizarre and decidedly anti-scientific notion that there is something significant (and comforting) in the fact that life from non-life cannot be demonstrated to be *impossible.*

The National Center for Science Education (NCSE), headed by atheistic biologist, Dr. Eugenie Scott, has for years been in the forefront of the battle to prevent the teaching of flaws in evolutionary theory or Intelligent Design theory in US public schools. Dr. Frank Sonleitner, a professor of zoology at the University of Oklahoma, has written a lengthy essay on the Origin of Life that appears on the NCSE website. He writes as follows:

> Modern ideas about the [emergence] of living things from non-living components...may not have yet come anywhere near answering all our questions about the process, but...none of this research has indicated that abiogenesis is **impossible**.[5]

To put Dr. Sonleitner's statement in a bit of perspective:

> None of our research, observations, or consultations with astronomers have come anywhere near answering our questions about the location of the china teapot, but nothing we have found in this investigation has indicated that Russell's "china-teapot-in-space theory" is **impossible**.

Dr. Paul Davies makes an almost identical declaration: "Just because scientists are uncertain how life began does not mean that life *cannot* have had a natural origin,"[6] (i.e., it's not **impossible**). Dr. Davies, please pay attention. It would have been at least as reasonable to state the following: "Scientists are absolutely stymied in their attempts to find a naturalistic origin of life and at the same time there is nothing to indicate that life *cannot* have begun through an act of Divine creation." Or alternatively: "Just because astronomers are uncertain does not mean that the china teapot ***cannot*** be in orbit." Even Francis Crick, undoubtedly one of the greatest scientific minds of the twentieth century, is not immune:

> An honest man, armed with all the knowledge available to us now, could only state that in some sense, the origin of life appears at the moment to be almost a *miracle*, so many are the conditions which would have had to have been satisfied to get it going. But this should not be taken to imply that...it **could not** have started on the earth by a perfectly reasonable sequence of fairly ordinary chemical reactions.[7] (In other words, it's not **impossible**.)

The following illustrates the absurd juxtaposition that Crick has presented us in his contradictory descriptions of the Origin of Life: Imagine winning 200 hands of black-jack in a row at a Las Vegas casino. As the pit-boss and his crew are summarily throwing you out of the casino onto the sidewalk for obviously cheating, you offer the following brilliant pleading: "I know it would be almost a *miracle* that I could win 200 hands in a row by pure luck, but it's not *impossible*!" Mark Isaak, in his book *The Counter-Creationism Handbook*, makes this very claim: "Nobody denies that the Origin of Life is an extremely difficult problem, that is has not been solved though does not mean that it is *impossible*."[8] Isaak then takes this fatally flawed reasoning to its ultimate level. In his section on Origin of Life, after listing six of the unproven speculative theories about the origin of life, he lists as number seven, and I'm not kidding: "something that no one has thought of yet."[9] Yes Mr. Isaak, "something that no one has thought of yet" is always a possibility and *never impossible*.

Bertrand Russell: Mr. Isaak, you claim to have evidence that the teapot really is in orbit.

Mark Isaak: Yes, Mr. Russell.

Bertrand Russell: What is your evidence, Mr. Isaak?

Mark Isaak: Something that no one has thought of yet.

Mark Isaak, Criminal Defense Lawyer: "Ladies and gentlemen of the jury, the District Attorney has presented eye-witness evidence against my client, fingerprint evidence against my client, and DNA evidence against my client. I would like to present a reason why you should find my client not guilty...uhhh...ummm...I've got it! Something that no one has thought of yet!"

Mark Isaak, Prosecuting Attorney: "Ladies and gentleman of the jury, I know I have not presented any evidence that the defendant is guilty, but no one has yet proved that it is **impossible** for him to be guilty!"

I don't know how to demonstrate that it's *impossible* for life to come from non-life any more than Richard Dawkins or Jerry Coyne know how to demonstrate that it's *impossible* for a china teapot to be revolving around the sun in an elliptical orbit between the Earth and Mars. But no rational person is going to believe either of those proposals without rock-solid evidence, and by the way, if we are accepting "it's not impossible" as an argument, how about the following: "it's not *impossible* that God created the world in six days and made it *look* like it's fourteen billion years old."

When the atheist says "it's *possible* that it happened," or "it's not *impossible* that it happened," or it could have been "something that no one has thought of yet," he is appealing to the notion of Infinite Possibilities. As we know from the courtroom, and as articulated so clearly by Bertrand Russell, we don't live in a world where we consider infinite possibilities, we live in a world where we consider *reasonable* possibilities.

As strange as it may seem, though, Dr. Eugene Koonin, a highly accomplished microbiologist — the same Eugene Koonin who informed us that Origin of Life research is an utter failure — has explicitly proposed that we should start considering infinite possibilities as a solution and escape hatch from having to accept a Creator of life. He writes the following:

> The Many Worlds in One version of the cosmological model of eternal inflation might suggest a way out of the origin of life conundrum because, in an infinite multiverse with a finite number of macroscopic histories (each repeated an infinite number of times), the emergence of even highly complex systems by chance is not just possible, but inevitable.[10]

Translation: The odds of rolling a six one thousand times in a row with a single die is one in 6^{1000}, or one chance in 10^{778}. The size of this number is beyond comprehension, but to provide some kind of baseline, keep in mind that the number of atoms in the entire universe is estimated to be roughly 10^{80}. Despite this, as Koonin points out, if I am able to roll the die an *infinite number of times*, it is not only possible, but *inevitable* that it will happen. Although reason and scientific investigation have informed us of the virtual impossibility of life having formed on our planet by an undirected naturalistic process, the "way out of the origin of life conundrum" — that is to say, the way to avoid the obvious answer that life was created — is to propose a multiverse. With an infinite number of trials and errors available, it is not only possible but inevitable that life will form no matter how fantastic the odds against.

He is, of course, correct. With an infinite number of trials and errors, not only is the formation of life inevitable but it is *just as inevitable* that any number of each of the following has formed by *pure chance* and can be found on our planet today: iPhone 5, Toshiba Satellite Laptop Computer, Schwinn Discover Men's Hybrid Bike, full color poster of Jimmy Hendrix playing at Woodstock, Martin D-35 Acoustic Guitar, Mylec Eclipse Jet-Flow Hockey Stick, Revell 1:48 scale P-51D Mustang model airplane, and last but not least, a 2014 Rolls Royce Phantom Sedan (retail price- $465,000). I don't believe it, Richard Dawkins doesn't believe it, Dr. Leonard Susskind of Stanford University doesn't believe it, Eugene Koonin himself doesn't believe it, and no one reading this book believes it.

Just as it is beyond absurd to propose that the Smartphone I am holding in my hand could be the product of chance, it is exponentially beyond absurd to propose that a bacterium could have formed by chance. There is either a flaw in Koonin's logic (which of course there is) or the multiverse theory is false and/or irrelevant to our question (which of course it is).

What we have seen then is that in the infinite space, or if you will the infinite *gap*, created by an infinite number of possibilities, there is plenty of room for the atheist to believe that life can come from non-life through some mysterious unguided process. It is there, in that infinite gap, that he finds a comfortable place to pitch his tent and call it home. Hence the Argument from Infinite Possibilities, or most appropriately of all, Atheism of the Gap.

Who Created the Creator, Who Designed the Designer?

To reiterate the point I made earlier, the public seems unaware that most serious atheistic thinkers reject the idea of an Intelligent Creator of life not because they have a scientific alternative (there is none) but because they claim it is *philosophically* untenable. In the minds of many prominent atheist polemicists, the question of "who designed the designer" presents the theist with a philosophical barrier so formidable that it cannot be breached. The late Christopher Hitchens raised this dilemma in his lecture at Sewanee University. When someone in the audience asked, "Where is the first cause, how can you do without a first cause," he replied: "Because it only gives you a sterile infinite regression. Where did the first cause of the first cause come from? The argument from design gives you the same problem; who designed the designer?"[11]

Jason Rosenhouse, a mathematician at James Madison University, elaborates:

> "Proponents of Intelligent Design [assert] that living organisms exhibit a certain kind of complexity...that is most plausibly explained as the result of intelligent design...the complexity of [the simplest living bacterium] is used as the evidence that a certain sort of designer exists."[12]

Rosenhouse points out what seems to be the inherent problem in proposing such a solution:

> This leads to a problem. The existence of complex entities was

> precisely the phenomenon in need of explanation. Hypothesizing the existence of something more complex than the thing to be explained only replaces one problem with a far greater one. If [the first living bacterium] can only be explained as the product of design, then any designer capable of crafting the [first living bacterium] must also be so explained. The result is an infinite regress of designers, each invoked to explain the existence of the one before.[13]

This philosophical question is actually the centerpiece of Richard Dawkins' rejection of Intelligent Design in *The God Delusion*:

> Seen clearly, intelligent design will turn out to be a redoubling of the problem. Once again, this is because the designer himself immediately raises the bigger problem of his own origin...any entity capable of designing something as improbable as [the first living bacterium] would have to be even more improbable than [the bacterium itself.][14]

To summarize: If it is highly unlikely that a bacterium could emerge without a designer, then it is even more unlikely that the designer of the bacterium could emerge without a designer, and so on. We then end up with the absurdity of an infinite regression of designers. For a number of different reasons, not the least of which is that we have accepted a *beginning* to the universe, which inherently precludes an infinite regress, no one is prepared to accept an infinite regress of designers/creators.

Before exploring the possibility of providing a solution to this dilemma, I must point out that this argument is routinely misunderstood and misapplied. The reader must understand clearly that the question of "who designed the designer?" has *no bearing or relevance whatsoever* in determining whether the bacterium, or anything else for that matter, is the result of intelligent causation. If I am looking at a cell phone, discover drawings in a cave in Spain, or receive intelligible radio transmissions from outer space, I immediately determine that they are caused by intelligent activity. But who designed the designer? I don't know, but that in no way affects my conclusion that there is intelligent activity involved. The nanotechnology of a bacterium is clearly the result of conscious, creative, intelligent activity. *But who designed the designer*? I don't know, but that doesn't change my conclusion.

The *only* implication that can be drawn from the question of "who designed the designer?" is that there cannot be an *infinite regression* of designers. There could be a regress of ten, fifty, or a thousand designers, there just cannot be an *infinite* regress. Let us put it a different way: the challenge of "who designed the designer" is not about determining whether or not the bacterium is the result of intelligent causation; it is *obviously* the result of intelligent causation. The challenge is ultimately whether or not we can discover a designer before whom there is no other designer...and as we shall see, we most certainly can.

Two Possible Beginnings

The one thing upon which both theists and their opponents agree is that there was a *beginning* to life. At one point in time life did not exist and then at some later point in time life did exist. There are only two possible beginnings. One, of course, is an unguided, undirected series of causes and effects entirely within the laws of chemistry and physics that resulted in the formation of life.

What is the other possible beginning? Imagine we have gone back in time and are looking at the very first living cell to ever exist. Somewhere, at some time, this *very first cell* had to be. The *truth* is that it was either created/designed or it was not. If the truth is that it was designed, what do we know about the designer? It is obvious that the designer could not possibly be physical; after all, this being the first living cell, there is no other physical being in the entire universe. This designer cannot be subject to the laws of cause and effect in time, because there cannot be an infinite regress, therefore this designer must be one before whom there is no other designer. In other words, a designer who is not physical and *does not exist in time*, for in that case there is no designer before this one, because there is no "before." As Dr. Paul Davies put it: "What happened before the Big Bang? The answer is there was no 'before.' Time itself began at the Big Bang." That is to say, a designer who is composed of neither matter nor energy and does not exist in time or space. *If* that designer exists, we are free from the dilemma of the infinite regress and the question of "who designed the designer." *If* the truth is that the first living cell was created/designed, it must be a designer/creator who fits this description. *If* we conclude that it is highly likely that the truth is that first living cell was designed, it inextricably follows that it is also highly likely that the truth is that this designer/creator whom we just described exists.

These are the two possible beginnings. The task before us is now very simple. All we must do is examine the evidence and *rationally* determine which beginning is more likely to be the *truth*. In other words, is it more likely that the first living cell was the result of intelligent causation or naturalistic, unguided forces? That question has already been dealt with in the last three chapters. Dr. Harold Urey, Nobel Prize-winning chemist, said the following in 1962: "All of us who study the origin of life find that the more we look into it, the more we feel it is too complex to have evolved anywhere. We all believe as an *article of faith* that life evolved from dead matter on this planet. It is just that its complexity is so great, it is hard for us to imagine that it did." Jacque Monod, Nobel Prize-winning biochemist, said that the chances of life emerging naturally were "virtually zero," but luckily, "our number came up in the Monte Carlo game." Dr. Urey was correct; you need to be a man of great faith indeed to believe that. How appropriate is the following observation made by one of my mentors, Rabbi Yaakov Weinberg,* of blessed memory, "If you close your eyes and take a leap of faith, you can land *anywhere you want to.*"[15]

I would suggest to the reader that not only is a supra-natural creator the only reasonable, rational, and logical solution to our question, but that we are "hard wired" to both understand and *experience* the reality of this concept. It is part of our inner essence. In October of 2007, the Fixed Point Foundation sponsored a debate at the University of Alabama between Richard Dawkins and John Lennox, a Christian mathematician from Oxford University. One particularly captivating moment in the

---

* Rabbi Yaakov Weinberg (1923–1999), formerly Dean of the Ner Israel Rabbinical Seminary in Baltimore, Maryland, was considered to have been among the foremost Talmudic scholars and theologians in the English-speaking Orthodox Jewish world.

debate was when, in my opinion, Richard Dawkins himself expressed this inner reality, and was unable to contain this genuine reaction to the wonders of the living world. He declared the following:

> I think that when you consider the beauty of the world and you wonder how it came to be what it is, you are naturally overwhelmed with a feeling of awe, a feeling of admiration...and you almost feel a desire to worship something... I feel this...I recognize that other scientists such as Carl Sagan feel this, Einstein felt it, we all of us share a common kind of religious reverence for the beauties of the universe, for the complexity of life, for the sheer magnitude of the cosmos, for the sheer magnitude of geological time...and it's tempting to translate that feeling of awe and worship...into a desire to worship some particular thing, a person, an agent...you want to attribute it to a Maker, to a Creator.[16]

What is Dawkins describing if not the experience of a reality that is indescribably greater than ourselves that transcends our own being? That the beauties of the universe, the unfathomable complexity of life, the sheer magnitude of the cosmos, inspire in us a desire to reach out somehow and connect with that ultimate greatness. A desire to not just connect, but a primal understanding that this ultimate "maker" is a being we want to *worship*. What stops him from taking that step? In Dawkins' own words:

> What science has now achieved is an emancipation from that impulse to attribute these things to a Creator... It was a supreme achievement of the human intellect to realize there is a better explanation...that these things can come about by purely natural causes...**we understand essentially how life came into being.**[17]

"We understand essentially how life came into being"?! Not true, Professor Dawkins; we don't even understand *non-essentially* how life came into being. Science is vexed, clueless, stymied, and baffled in its attempt to understand a naturalistic origin of life. How could you possibly make such a disingenuous statement when you yourself know this to be the true state of affairs? Here is a verbatim transcript of an interview of

Richard Dawkins by Ben Stein in the 2008 documentary film, *Expelled: No Intelligence Allowed*:

Stein: How did it start?

Dawkins: *Nobody knows how it started*, we know the kind of event that it must have been, we know the sort of event that must have happened for the origin of life.

Stein: What was that?

Dawkins: It was the origin of the first self-replicating molecule.

Stein: How did that happen?

Dawkins: *I told you I don't know.*

Stein: So you have no idea how it started?

Dawkins: *No, no, nor does anyone else.*[18]

When Dawkins told his overflow audience that science understood "essentially" how life came into being, was he simply fudging to score points in the debate? Deluding himself? Deliberately lying? I have no way of knowing, but I do know that what he said was utterly false. Science has most definitely not emancipated us from the impulse to attribute these things to a Creator nor has any human intellect proffered anything remotely resembling the "supreme achievement" of explaining the Origin of Life in natural terms. Only this profound intellectual/psychological disconnect, this unmistakable example of cognitive dissonance, allows Dawkins to continue to deny the existence of a Creator of life.

Perhaps it is time for Dawkins and his colleagues to express not awe, not admiration, but *humility*. To stand back from their lab tables, test tubes and Bunsen burners, and reflect on the words of the Psalmist: "My heart was not proud, and my eyes were not haughty, nor did I pursue matters too great and wondrous for me..." (Psalms 131)

Professor Dawkins stated in his debate with Dr. Lennox, "[I]t is *tempting* to translate that feeling...into a desire to worship some particular thing... you want to attribute it to a maker, a creator." My advice to Dawkins and his colleagues? Don't go to church, don't go to synagogue, don't plan a trip to Mecca, India, Nepal, or for that matter to Salt Lake City. Just do it, give in to temptation. And if you find yourself unable to kneel and worship, at the very least stand bowed in humility before your Creator.

End Notes

1 Dr. Jerry Coyne, "A Rabbi Proves God."
2 Dawkins, *The Blind Watchmaker*, p. 5.
3 Hitchens, *God Is Not Great*, p. 65.
4 Bertrand Russell, see http://en.wikipedia.org/wiki/Russell%27s_teapot
5 From the NCSE website: http://ncseweb.org/creationism/analysis/excursion-chapter-1-origin-life
6 Davies, *The Fifth Miracle*, p. 31.
7 Crick, *Life Itself*, p. 88.
8 Mark Isaak, *The Counter-Creationism Handbook* (Los Angeles, CA: University of California Press, 2007), p. 45.
9 Ibid.
10 Koonin, *The Logic of Chance*, p. 392.
11 "Why Atheism is Necessary for Morality," Hitchens lecture at Sewanee University, YouTube.com
12 Jason Rosenhouse, "Who Designed the Designer?" — http://www.csicop.org/specialarticles/show/who_designed_the_designer
13 Ibid.
14 Dawkins, *The God Delusion*, pp. 145–146.
15 Heard by the author in a lecture by Rabbi Yaakov Weinberg, of blessed memory.
16 Lennox-Dawkins debate, www.dawkinslennoxdebate.com
17 Ibid.
18 *Expelled: No Intelligence Allowed*, 2007 film, http://www.youtube.com/watch?v=BoncJBrrdQ8

Section 3

Man's Search for Meaning and Spirituality

Chapter 6

The God that We are Seeking

The illustrious British astronomer, physicist, and mathematician, Sir Fred Hoyle (1915–2001) wrote that "a common sense interpretation of the facts suggests that a super intellect has monkeyed with physics, as well as with chemistry and biology, and that there are no blind forces worth speaking about in nature."[1] While a description of God as the "super intellect" behind physics, biology, and chemistry, is no small thing, I propose to the reader that God is much more to us than a master Engineer who is the creator/designer of the machinery of life. I would suggest that our whole purpose in life revolves around building a relationship with this God. That is to say, he is the God whom *we are seeking.*

Over the next few chapters we will explore and develop this theme by focusing on three fundamental philosophical and metaphysical issues that believers constantly point to in their ongoing debate with skeptics and non-theists:

1. Man's search for meaning
2. The seemingly non-material (i.e., spiritual) realities that permeate our lives (for example, the mind/brain duality)
3. Morality

At first glance, each of these concepts could suggest a reality that extends beyond the material universe; a reality that exists in time but not space; a reality that profoundly affects our lives but cannot be quantified or detected by any conceivable material yardstick or measuring device. **In fact, this is to be our working definition of the spiritual**: a reality that exists in time but not space, has a clear effect on our lives, and is undetectable and un-measureable by physical means. When confronting these issues, many skeptics and freethinkers tend to view themselves as having a virtual monopoly on the commodities of rational discourse and contemplation. The following passage from Dr. Julian Baggini's book, *Atheism: a Very Short Introduction*, conveys the attitude just described: "Most atheists see themselves as realists — their atheism is a part of their willingness to square up to the world as it is and face it without recourse to superstitions or comforting fictions about a life to come or a benevolent power looking after us."[2]

I agree that atheists do not generally resort to actual superstitions in order to "square up to the world as it is." However, I would suggest that in order to keep their non-belief intact, they routinely *do* turn to "comforting fictions." I contend that a careful analysis of these "comforting fictions" will lead us to an extraordinary insight. We will see that in his very remonstrations against religious belief, the atheist in fact testifies that the deepest drives and impulses in the human psyche can only be understood as functions of the inner need to seek and relate to a transcendent God. Paradoxically then, over the next few chapters we will use atheistic philosophy itself as a sort of radio beacon to navigate a course to a rendezvous with our Creator. To begin the journey, we present the musings of Richard Dawkins on the meaning and purpose of life.

Dawkins and the Meaning of Life

In an article written for *Scientific American* (1995), Dawkins informs us in blunt, raw language his existential view of reality: "The universe that we observe has precisely the properties we should expect if there is, at bottom, no design, no purpose, no evil, no good, nothing but pitiless indifference."[3]

I don't think I would be going out on a limb by opining that the universe described in the above statement is not a universe that would inspire the average person to jump for joy (it might however inspire him to jump off a tall building). How does a prominent atheist like Richard Dawkins deal on an emotional and psychological level with what is ostensibly a horribly depressing and despair-inducing view of reality? At first it would seem as if Dawkins is exemplifying the ideal described by Baggini. He is willing to "square up to the world as it is" without seeking comfort in imaginary constructs. He admits the simple truth — that a Godless world is one that has no purpose or value, only "pitiless indifference." But that still does not answer the question I posed above. How does he deal with it? In fact, he deals with it as most non-theists deal with it: by rushing headlong into the embrace of a comforting fiction.

> Isn't it a **noble**, an **enlightened** way of spending our brief time in the sun, to work at understanding the universe, and how we have come to wake up in it?... Isn't it sad to go to your grave without ever wondering **why you were born**? Who, with such a thought, would not spring from bed, eager to resume discovering the world and rejoicing to be a part of it?[4]

Noble — of an exalted moral or mental character (*Random House Dictionary, 2010*). **Enlightened** — having knowledge and spiritual insight, characterized by full comprehension of the problem involved (*American Heritage Dictionary of the English Language, 2006*).

How is it possible to say in one breath that our universe has "no design, no purpose...nothing but pitiless indifference," which explicitly means that the universe is devoid of all significance, and a moment later describe devoting one's life to understanding that *same* universe as a "noble" and "enlightened" way of life, adjectives that are literally bursting with implications of significance? Why would anyone be "eager to resume discovering [a] world, and rejoicing [to be part of a world]" that has neither design nor purpose?

What does it mean when we say something has no purpose? A water bucket whose bottom is full of holes has no purpose. Would anyone dream of examining or trying to understand a bucket that is full of holes or would you simply throw it in the garbage? Imagine assigning a worker the task of filling up such a bucket with water. How long before he would throw the bucket against the wall cursing and walk away?

Let's focus on a skycap at the airport whose job essentially consists of putting tags on luggage and then placing that tagged luggage on a conveyor belt. It may not be a job that requires advanced education, but it is certainly an honest and honorable way to support one's family and earn a living. He can work for an airline for thirty-five years, retire, and live off his pension. Take the same skycap, tell him to put tags on luggage, put the luggage on a conveyor belt that goes around in a circle, and when the luggage comes back around his job is to take it off and start all over again. Can you imagine doing *that* for eight hours a day for thirty-five years?! How long would it take before he would start howling at the sky like a raving lunatic?

What is the difference between the two scenarios? In the first, the skycap feels there is some *purpose and meaning* in what he is doing. People travel to different locations for business and pleasure. They also need their luggage transported to their destination and the skycap is part of this process. In the second scenario, he is doing the *same work*, but it is purposeless and meaningless. An ox can turn a millstone for its entire life without the slightest awareness or concern if there is grain under the millstone or not, as long as he is fed and gets adequate rest. Put a human being to work at a purposeless task for a long enough period of time and he will most likely go stark-raving mad.

A survivor of the Soviet Gulag once described how inmates were marched out in the dead of winter and told to dig a long narrow ditch, ostensibly for some sort of piping. The ground of course was frozen solid and the work was excruciatingly difficult. The inmates returned to the prison camp exhausted beyond description. The next morning they were marched to the same spot and were ordered to fill in the ditch with the same dirt. Which was easier work — digging the ditch or filling it in? Which was more painful? To order sick and malnourished men to dig a ditch in the middle of a Siberian winter will break men's bodies, but to have them fill it in the next day and to realize that all that backbreaking work was absolutely purposeless will crush men's spirits and souls (which of course is exactly what it was designed to do).

This then is the authentic human reaction when confronted with purposelessness, whether it is in the form of filling a hole-riddled bucket with water, putting luggage on a conveyor belt going nowhere, or filling in a ditch with dirt. When faced with a purposeless situation, human beings do not see it as *noble and enlightened*; they do not approach it *eagerly and joyfully*. Purposelessness is a state of being that is unbearably painful for the human being. People who ***truly*** feel that life is purposeless will either take drastic steps to block out that reality and the accompanying pain (e.g., with drugs, alcohol, etc., or a "comforting fiction") or when despairing of a solution will end the agony with suicide.

A universe with "no design, no purpose, no evil, no good, nothing but pitiless indifference," is essentially one giant cosmic conveyor belt going around in meaningless circles. In "Mr. Dawkins' Neighborhood," we are all a bunch of skycaps trapped in a Twilight Zone-like surrealistic horror film. The universe of Richard Dawkins is one that offers a life filled with agonizing despair. On the other hand, a life that is "noble and enlightened," a life that is approached with "joy and eagerness," is a life *rich with purpose and meaning*. It is the type of life that can only be lived in a universe that has been created with ultimate goals and ultimate purposes. Does Richard Dawkins actually believe that he lives in a meaningless world and then psychologically and emotionally covers it up with his "comforting fiction," which consists of a fantasy about the nobility of the pursuit of scientific knowledge, or in his heart does he really believe there is an ultimate purpose to our lives and simply avoids having to confront his Creator by calling himself an atheist?

"Isn't It Sad to Go to Your Grave Without Ever Wondering Why You Were Born?"

Is it actually possible that the above question was propounded by Richard Dawkins, perhaps the most articulate, outspoken, passionate, and loyal disciple of Charles Darwin alive today? The late Dr. Stephen J.

Gould, world-renowned paleontologist, has already told us the obvious answer to this question:

> We are here because one odd group of fishes had a peculiar fin anatomy that could transform into legs for terrestrial creatures; because the earth never froze entirely during the ice age; because a small and tenuous species, arising in Africa a quarter of a million years ago, has managed so far to survive by hook and by crook. We may yearn for a higher answer — but none exists.[5]

In other words, Professor Dawkins, you were born for *no reason at all*. You were born because of a meaningless series of coincidences and happenstance. You are a flaming accident! You started from nothing and you are going nowhere. Is Richard Dawkins a Darwinist or some Eastern mystic seeking to unravel the mystery of the universe and our existence?

How are we to understand this philosophical version of a Dr. Jekyll/Mr. Hyde-like dichotomy in the world outlook of Richard Dawkins? On the one hand the non-believing, existentialist philosopher who sees a purposeless, insignificant, indifferent, pitiless world, and on the other hand the bedazzled awestruck child, the mystical seeker who yearns to understand the universe, why we are here, and what is our unique place in it? Is there any way of harmonizing these two views of reality? Let's examine how this seemingly irresolvable tension expresses itself among other atheist writers and philosophers.

Sigmund Freud: "The moment a man questions the meaning and value of life he is sick, since objectively neither has any existence."[6]

Christopher Hitchens: "Why care? Why do I bother? That's a very good question. It...doesn't have a conclusive answer."[7]

Steven Weinberg: "The more we know of the cosmos, the more meaningless it appears."[8]

Richard Dawkins: "The universe...has...no design, no purpose...nothing but pitiless indifference."[9]

Dr. James Watson: "I don't think we're here for anything, we're just products of evolution. You can say, 'gee, your life must be pretty bleak if you don't think there is a purpose,' but I'm anticipating having a good lunch."[10] [Alimentary, my dear Watson...]

Stephen J. Gould: "We may yearn for a higher answer — but none exists."[11]

Jean-Paul Sartre: "Life has no meaning the moment you lose the illusion of being eternal."[12]

Carl Sagan: "The very scale of the universe....speaks to us of the inconsequentiality of human events in the cosmic context."[13]

G. Gaylord Simpson: "Man is the result of a purposeless...evolutionary process..."[14]

Emile Cioran (Romanian philosopher): "I'm simply an accident, why take it all so seriously?"[15]

Will Provine: "There is no hope whatsoever in there being any deeper meaning in life."[16]

To sum up: *All meaning in life is based on an illusion. There is no compelling reason to care about anything, no reason to bother with anything; we are here by accident; there is no reason to take it seriously. We live in a purposeless, insignificant, indifferent, pointless universe. Life has no objective value or meaning; our deeds and doings are inconsequential, we are not here for anything at all, there is no hope of finding deeper meaning in life, we are products of a blind and directionless evolutionary process, and it makes a man sick to think about it. We may yearn for a higher answer — but none exists.*

I did not make these statements, *they did.* All of them, in one way or another, exhibit the same schizoid behavior that we pointed out above in Richard Dawkins. Why are these people writing books? Why are they giving lectures? Why do they all seek achievement? Why are they chasing awards and honors (other than to give Steven Weinberg's mother the chance to say, "My son, the Nobel Prize winner!")? There is no logical or rational reason for the atheist to do any of these things. Pointless means *pointless.* It means that the lectures don't matter; it means the books are inconsequential; it means the values hold no significance — there is *no point* to it. Jean-Paul Sartre, the French philosopher and author, spent his entire lifetime writing books explaining why there is really no meaning to writing books; giving lectures about how there is no real point to giving lectures, and teaching students that there is no real purpose in teaching students. His passion to succeed and achieve is an object lesson in theater of the absurd.

When Sigmund Freud awoke in the morning and looked in the mirror, did he say, "Remember, Sigmund, don't think about the value and purpose of your life, because in reality it has none. If you keep thinking about it, you are going to need a psychiatrist!" Do you think I am gratuitously mocking Freud? I wasn't the one who said that when a man thinks about the meaning and value of life he becomes sick because in fact there is no meaning and value to life — *Freud said it*.

Perhaps the following is the most glaring question of all for the atheist ideologue: For what possible reason would you care if people believe in God and religion or not? Dr. Steven Weinberg, a brilliant physicist, has declared that, "The more we know of the cosmos, the more meaningless it appears."[17] He then turns around and says, "Anything that we scientists can do to weaken the hold of religion should be done, and may in the end be our greatest contribution to civilization."[18]

What upsets Weinberg about religion? If Richard Dawkins is allowed his comforting fiction, why can't religious people have their comforting fiction? It certainly is not because life is necessarily more pleasant without religion. Many people are very happy living a religious lifestyle. The simple historical fact is that atheist tyrants and their societies have been *at least* as tyrannical, vicious, and destructive as any repressive religious society. How about North Korea, the former Soviet Union, Red China, the former East Germany, Cambodia under the Khmer Rouge, North Vietnam, and Cuba? In fact, many scientists *themselves* have been guilty of the most horrible atrocities in recent history.

Is it because religion teaches that there *is* a meaning to the universe and Weinberg demands that everyone admit that it's really meaningless? If it's meaningless, then what is the point to caring what anyone believes? What does he mean by "our greatest contribution to civilization"? That everyone should agree to the atheist doctrine that civilization itself is pointless and meaningless? Is this the great intellectual and ideological victory that Weinberg dreams of? This type of thinking borders on madness.

But perhaps I have misunderstood. Perhaps the *universe* is pointless and meaningless, but we can fashion our own meaning for *human civilization.*

Can We Create Our Own Meaning?

Atheistic philosophers tell us that the only escape from the aforementioned dilemma is to create our own *subjective* meaning and purpose in life:

"Human life has no meaning independent of itself...the meaning of life is what we choose to give it."[19] (Paul Kurtz, Humanist philosopher)

"Life has no meaning *a priori*...it's up to you to give it a meaning, and value is nothing else but the meaning you choose...man is alone... abandoned on earth...with no other aim than the one he sets himself."[20] (Jean-Paul Sartre)

Before analyzing what Kurtz and Sartre have proposed, it is instructive to actually define what "subjective" means: "proceeding from or taking place in a person's mind rather than the external world...existing only within the experiencer's mind...existing only in the mind, illusory" (*American Heritage Dictionary*).

The idea of creating our own subjective meaning and purpose in life may sound very profound in the university lecture hall or in some late night university-dorm rap session. However, when stripped of its philosophical camouflage, what it really means is the following: Make something up that gives your life purpose and pretend that it's real. Create a fantasy world, an illusion in your mind so that you will not, in the words of novelist T. C. Boyle, have to face the "naked howling face of the universe" and live in existential terror.

We have now come full circle from the statement of Julian Baggini that was cited at the beginning of this chapter: "Their atheism is part of their willingness to square up to the world as it is...without recourse [to] comforting fictions...." It has now become crystal clear that what the atheist philosopher truly preaches is exactly the opposite. In other words, do *not* "square up to the [meaningless] world as it is." In other words, hold on tight to a "comforting fiction." In other words, hook yourself up to "The Matrix" and attempt to live a blissful hallucination. The sum total of all the anguished, harrowing probing of the human psyche by philosophers like Sartre and Kurtz is really nothing more than an active expression of Freud's observation — that is to say, figure out a way to avoid confronting the absolute meaningless and valueless reality of our existence lest, as Freud put it, you become "sick" (or decide to take a long walk off a short ledge).

In *God: The Failed Hypothesis*, physicist Victor Stenger confronts this dilemma of the ages head on. Stenger admits that in the purely material world of the non-believer, ultimately nothing we do matters. However, he offers what he believes is an ingenious solution to the problem:

> Philosopher Erik Wielenberg tells of a gym teacher who would calm things down when tempers flared...by saying, "Ten years from now, will any of you care who won this game?" Wielenberg recalls thinking that a reasonable response would be: "Does it really matter **now** whether any of us will care in ten years?" He quotes philosopher Thomas Nagel in the same vein, "It does not matter now that in a million years nothing we do now will matter." In other words, what matters now is what happens now.[21]

Let's expand on Stenger's penetrating insight that "what matters *now*, is what happens *now*":

- It doesn't matter *now* that I haven't attended any classes all semester, haven't opened a book, and will be expelled from school; what matters *now* is that I am partying and having a great time.
- It doesn't matter *now* that I'm taking heroin and will become a broken down junkie within months; what matters *now* is that I am on a high.
- It doesn't matter *now* that if I commit murder I might end up in jail for the rest of my life; what matters *now* is that I am killing the liquor store owner and getting $1,000 out of the cash register.
- It doesn't matter *now* that, as you face the end of your life *like all of us will one day*, you might be confronted with a paralyzing thought: **What was the point to all this? What difference did it all make**? After all, if you don't think about it, you can pretend the question doesn't exist.

While in the midst of a tantrum, a five-year-old is not expected to see beyond right now. However, one of the characteristics that distinguishes adults from children is the ability to understand the consequences and implications of their ideas, beliefs, and actions. What is Stenger's solution to the profound existential dilemma that the greatest intellects in human history have grappled with for millennia? *It doesn't matter that in objective reality everything you are doing is insignificant and meaningless. Don't think about that, push it out of your mind. What difference does it make how you will feel about your life in five years, or ten years, or twenty years? The only thing that matters is* ***now****!* In fact Stenger's brilliant new philosophy is really a very old philosophy: "Eat, drink, and be merry, for tomorrow we die." Thank you, Victor Stenger, for sharing; your particular "comforting fiction" is *very special*!

When addressing the meaning of life, Sartre, Dawkins, Weinberg, Freud, etc., bear an unmistakable resemblance to a group of dogs going around in circles chasing their proverbial tails. The universe of Dawkins is purposeless ("the universe...has no design...no purpose"), but he finds great purpose in studying it and writing one book after the other about it. The universe of Dr. Steven Weinberg is meaningless ("the more we know of the cosmos, the more meaningless it appears"), but he finds great meaning in devoting his life to investigating it and winning a Nobel Prize in the process. In Freud's evaluation human life has no value ("the moment a man questions the meaning and *value of life* he is sick, since objectively neither has any existence"), but he found it immensely valuable to study human behavior and publish prolifically on the subject. It is clear that these men are driven to define their lives in the context of some meaningful and purposeful activity, even though their espoused atheistic ideology explicitly denies, as Dr. Will Provine put it, even the "hope" of the existence of such a thing. They are unable to escape this relentless, inexorable drive. Why?

We have already pointed out that unlike the ox that goes in circles around the millstone and could care less about why or what he is accomplishing, the human being simply cannot live without purpose. The human need for meaning is as real and as critical as our need for oxygen, water, and food. It goes further than that. Our need for meaning is *greater* than our need for oxygen, water, and food. It is greater than our need for friendship, love, or happiness. Human beings will *sacrifice their lives* for what they believe is meaningful. This is not some abstract, intellectual concept; it is hardcore, everyday reality. People all over the world are dying, and are prepared to die, for what they believe in. It cuts across all doctrines and belief systems. It applies to religious ideologies and atheistic ideologies. Whatever a particular individual believes is the essential reason for and central meaning of his existence — whether it is to defend one's family, honor, country, or faith — that individual will be prepared to give up his or her life for it (at the very least, the potential is there for this individual to give up his or her life). What need could possibly be so powerful as to be more important than life itself?

Meaning and Purpose beyond Physical Existence

The meaning and purpose we are talking about here is not the utilitarian type of purpose to acquire food or fulfill some other practical necessity. That type of drive we share with all living things. We are talking about someone who has every physical need taken care of and still yearns for something "more."* It is the type of meaning that Sartre refers to when he says, "Life has no meaning the moment one loses the illusion of being eternal." It is the type of meaning that Freud talks about when he says that a man becomes sick when he realizes that there is no objective meaning to his life. It is the type of meaning that Holocaust survivor Dr. Victor Frankl** means when he writes: "What man actually needs is not a tensionless state, but rather the striving and struggling for some goal worthy of him. What he needs is...the call of a potential meaning waiting to be fulfilled by him."[22] It is the type of meaning human beings crave so powerfully; that a split second after an atheist declares human existence and the universe to be void of all significance, that very same

---

* Imagine a healthy, robust, well-fed ox turning a millstone and suddenly thinking to himself, "I want *more* from my life."

** Frankl was an Austrian neurologist and psychiatrist who created Logotherapy, a school of psychotherapy, and authored a famous chronicle of his life in the Nazi death camps entitled *Man's Search for Meaning.*

atheist is driven to declare his passionate commitment to investigating and understanding human existence and the universe. This burning hunger for meaning and purpose has no atomic structure, no molecular configuration, and no chemical formula with which it can be satiated. The only "food" that satisfies this ravenous desire is some transcendent goal or quest. It is this deep yearning that drives Sartre to exclaim, "That God does not exist I cannot deny, that my whole being cries out for God, I cannot forget."[23]

From Where Emanates Our Need for Transcendence?

Where does an avowed materialist and atheist like Sartre get the idea of a God that his "whole being cries out for?" What achingly profound inner need is at work for him to understand that if God existed, that's what he is looking for? How is it possible for an outspoken atheist like Sam Harris to make the following statement: "There is clearly a sacred dimension to our existence and coming to terms with it could well be the highest purpose of human life"?[24] What could the phrase "sacred dimension to our existence" possibly mean to a non-believer? For that matter, what could the phrase "higher purpose" mean to Sam Harris? We have already seen that prominent atheists declare to us that life has no purpose at all! What forces are churning inside the soul of Richard Dawkins that would produce these words: "When you consider the beauty of the world, and you wonder how it came to be...you are naturally overwhelmed with a feeling of awe, a feeling of admiration, and you almost feel a desire to worship something...you want to attribute it to a maker, a creator."[25] Dawkins, Harris, and Sartre speak for billions of other human beings when they express that their very being cries out to connect with something beyond this world, which transcends this world, to worship "a maker, a creator," to experience a "sacred dimension" and a "higher purpose" that would provide them with the fulfillment and significance that they feel is lacking.

Where is there room for such thoughts and concepts in a purely physical universe? In the atheistic view of reality, there isn't anything beyond this world, there is no such thing as transcendence. None of these things have any existence or reality. To propose an answer that primitive man dreamed up God or gods to explain and/or to achieve a sort of harmony or inner peace with the powerful and overwhelming forces of nature, is shamelessly begging the question. Why would primitive man, or for that matter modern man, even bother to think of such a question? From what part of a person springs the *need* to seek some ultimate explanation of himself and the world around him? Why does it eat at him and relentlessly drive him? If it doesn't bother zebras, dolphins, or chimpanzees, why should it bother us? Why can't people be happy and satisfied to live purely physical, material lives like the purely physical primates that we *actually are* in the atheistic worldview?

Imagine if the laws of physics and chemistry in our universe were slightly different and precluded the existence of the state of matter we call liquid. Could anyone conceive of a thing called liquid? Try right now to conceive of a state of matter that could not by definition exist in our universe. To attempt to do so would be ridiculous; you can't possibly have any idea of what you're trying to conceive of! It's like asking the blind-from-birth man to conceive of a flag made from red, white, and blue. Colors are simply not part of his reality. Actually we can take it a step further. Imagine if our entire universe only consisted of "black and white." *Nobody* would ever talk about or even dream about red, green, or blue. If the state of matter we call liquid did not exist, it would be absurd (in fact it would be impossible) to talk of a living creature being thirsty.

We cannot have a drive and need for something that has no actuality and existence. In a purely materialistic universe, there is no such thing as "purpose" or "meaning," only atoms, light waves, and the laws of physics. *They would simply and absolutely not have any place in the fabric of reality, including our thoughts and imagination.* In other words we would all be like the ox turning the millstone, perfectly content as long as our physical needs were taken care of. "Meaning" cannot be metamorphosed, transfigured, or expressed as a function of any material formula. It is just as absurd for a living creature to need meaning in an absolutely material world as it would be for that same creature to be thirsty if the laws of physics and chemistry did not allow for the existence of liquid. In this sense, the fact that I get thirsty incontrovertibly testifies to the existence of a state of matter called liquid. In the same way, my overwhelming "thirst" for meaning testifies to the existence of a thing called "meaning."

Our desire for meaning is real; we "cry out" for meaning. The urgent compulsion to seek meaning is more powerful than life itself; we will trade our lives for meaning. Meaning exists. The cosmos is not "meaningless," as Steven Weinberg put it. It is *not* true that "we're not here for anything." The universe has *purpose,* and we are most definitely here for *something.* We are inescapably hardwired to seek a meaning that transcends our physical existence. A search for meaning that ultimately can only find it's fulfillment in a connection to something "noble and enlightened," to a "sacred dimension," a "higher purpose," a "maker...a creator"....the eternal, One God. It is as futile to attempt to escape this implacable urge and the transcendent reality that implicitly must exist as it is to attempt to escape the need to eat and breathe.

As we have illustrated, our quest for meaning eloquently testifies to the existence of a higher realm; however, the concept of spirituality needs to be explored at a deeper level. Is it really possible to account for the entire spectrum of day-to-day human experience within the purely material paradigms of the atheistic conception of reality? Or is there perhaps...more?

End Notes

1 *Engineering and Science*, November 1981, pp. 8–12.
2 Julian Baggini, *Atheism: a Very Short Introduction*, (Oxford University Press, 2003), p. 10.
3 As cited in Stenger, *God: The Failed Hypotheses*, p. 71.
4 Ibid., p. 256.
5 Konner, *The Atheists Bible*, p. 2.
6 *Letters of Sigmund Freud*, p. 436.
7 Hitchens, "The Moral Necessity of Atheism," lecture at Sewanee University, YouTube.com
8 Cited in Dr. Stuart Kauffman, *Dreams of a Final Theory*, http://www.edge.org/3rd_culture/kauffman06/kauffman06_index.html
9 Stenger, *God: The Failed Hypotheses*, p. 71.
10 Dawkins, *God Delusion*, p. 126.
11 Konner, *The Atheists Bible*, p. 2.
12 http://thinkexist.com/quotation/life_has_no_meaning_the_moment_you_loose_the/210451.html

13 Sagan, *Broca's Brain*, p. 341.
14 Simpson, *The Meaning of Evolution*, p. 241.
15 http://en.wikipedia.org/wiki/Emil_Cioran; http://www.brainyquote.com/quotes/authors/e/emile_m_cioran_2.html
16 From an interview in the film *Expelled: No Intelligence Allowed*, http://www.youtube.com/watch?v=vuVSIG265b4&feature=related
17 Cited in Kauffman, *Dreams*, http://www.edge.org/3rd_culture/kauffman06/kauffman06_index.html
18 Konner, *The Atheist's Bible*, p. 2.
19 Ibid., p. 20.
20 http://thinkexist.com/quotes/like/life-has-no-meaning-a-priori-before-you-come/397354/ http://en.wikiquote.org/wiki/June_21
21 Stenger, *God: The Failed Hypotheses*, p. 251.
22 Victor Frankl, *Man's Search for Meaning* (Simon and Schuster, Inc., 1985), p. 127.
23 http://thinkexist.com/quotation/that_god_does_not_exist-i_cannot_deny-that_my/189882.html http://atheisme.free.fr/Quotes/Sartre.htm
24 Harris, *The End of Faith*, p. 16.
25 Lennox-Dawkins debate, www.dawkinslennoxdebate.com

Chapter 7

The World of Spirituality

In his book *Atheism: A Very Short Introduction*, Dr. Julian Baggini addresses a controversy that continually bedevils the atheist/materialist. What is the nature of "consciousness" and "self-awareness"? Are they manifestations of a non-material/spiritual reality or are they explainable and understandable as purely material phenomena? In his discussion of this subject, Baggini displays a disappointingly narrow-minded attitude:

> What best explains the correlation between consciousness and brain activity...[the] atheist hypothesis that consciousness is a product of brain activity or an **implausible tale** about how non-material souls exist alongside brains and somehow interact with them?[1]

The classic example of a declarative statement in the form of a question is "when did you stop beating your wife?" When Baggini queries about the "implausible" existence of the soul, he is clearly making a statement, not posing a question. Why did he feel it necessary to frame the issue in this manner? Why is it an "implausible tale" to consider the existence of a non-material soul?

Roget's II: The New Thesaurus lists the following as some of the synonyms for *implausible*: "flimsy, improbable, inconceivable, weak, unconvincing, thin, shaky." *Random House Unabridged Dictionary*: "not having the appearance of truth or credibility." *WordNet 3.0*: "having a quality that provokes disbelief," "a farfetched excuse."

If the viewpoint of the believer on this issue is "improbable," "not having the appearance of truth or credibility," "shaky," "weak," or in other words, if it is so obvious that the atheistic explanation is much more sensible, why not just pose the question in a neutral manner (as follows below), and let the reader come to the obvious conclusion:

> What best explains the correlation and relationship between consciousness or self-awareness, and brain activity? The atheist claims that consciousness is a product of brain activity and is wholly explainable as a material phenomenon. The believer claims that consciousness and self-awareness represent a non-material dimension of the soul that clearly interacts with the brain and our physical being, but still expresses a separate aspect of human experience and existence.

Once we read a neutral formulation of the question, the reason why Baggini felt compelled to add the words "implausible tale" becomes rather obvious. When posed neutrally, it is apparent that the suggestion that consciousness and self-awareness are manifestations of a separate non-material reality is not really "farfetched," "having a quality that provokes disbelief," "inconceivable," "flimsy," or implausible at all.

To cavalierly dismiss the notion of a non-material soul as implausible when billions of human beings from every conceivable race, culture, geographical location, and level of education claim an intuitive and experiential connection with its reality, is a flagrant display of intellectual laziness and pompousness. It is clear that no matter what conclusion is eventually reached, the honest thinker must consider both sides. Baggini of course, wants to avoid opening *that* can of worms. With his calculated fashioning of the question he eliminates the need to consider another point of view. Not only does he not offer any logical or scientific explanation for his *a priori* rejection of the soul and a corresponding non-material reality, but on the contrary, he explicitly informs us that this conclusion has been reached via a leap of faith:

> What most atheists do believe is that although there is only one kind of stuff in the universe and it is physical, out of this stuff comes minds, beauty, emotions, and moral values.[2]

"What most atheists do *believe*"?! Is it possible that a strictly rational, logical, skeptical, scientific atheist is using the "B" word? Do atheists have "beliefs"? I thought only medieval, backward, superstitious, religious people had "beliefs." Atheists deal only in scientific and empirically proven "facts." Is Baggini implying that he really does not *know* if the only "stuff" in the universe is physical and from that physical "stuff" comes minds (i.e., consciousness and self-awareness), beauty, and morality? Richard Dawkins cites this paragraph verbatim in *The God Delusion*.[3] It seems that Dawkins also does not really *know* if this statement is true or not. Praise Darwin! Baggini and Dawkins are *true believers*.

The simple truth is that of course they don't *know* if the only thing in the universe is physical "stuff." They have no idea whatsoever if love, beauty, and emotions can be defined in terms of some purely material configuration. If anyone knew the chemical or molecular formula for love, they would have patented it and become billionaires many times over. For them it is an *article of faith*. Once they have chosen to deny the existence of God, they have no choice but to also deny that there can be any type of reality besides one that is exclusively material. According to them nothing exists besides atoms, molecules, chemicals, light waves, and the laws of physics. Thus, beauty, emotions, and morality must also be defined in physical terms, no matter how counterintuitive or bizarre that seems.

I would like to suggest a completely different reason why it's possible for the skeptic to deny the actuality of the spiritual world. In the same way that a fish may never notice the water because it is completely immersed in it, it is also possible for people to fail to notice the spiritual side of life because our whole life and being is so completely immersed in a spiritual existence. Let's now take a little plunge into the ocean of our spiritual world...

The Spiritual Ocean

Who is "I"? Let's imagine for a moment that all the thoughts and pictures in our heads can be defined physically. Perhaps they are arrangements of electrons like the picture on a TV screen. This would mean that as I think and I see things inside my head, it's like watching a type of cerebral TV. The essential mystery that must be solved, however, is in the statement, "*I* am watching a cerebral TV screen." Who is doing the watching? No matter how we explain it, *somebody* is certainly watching *something*. Is the "I" also a configuration of electrons? Electrons watching electrons? Do the other electrons stare back?

The "I" is clearly separate from everything else going on in my head. At the very least, that is the way that every human being perceives it and until demonstrated otherwise there is no reason to doubt it. I have a brain, but "I" am not a brain. I have a body, but "I" am not a body. I feel emotions, but "I" am not an emotion. I think thoughts, but "I" am not a thought. Who is doing the thinking? Who is doing the feeling? Who is doing the perceiving? Who is this inescapable "I"? The existence of the "I" does not need to be proven. The "I," in fact, needs nothing at all to assert, justify, or confirm its own reality. The "I" simply *is*. It would seem that the classic Cartesian formulation, "I think therefore I am," would be more precisely stated as "I *am*, therefore I think;" "I *am*, therefore I feel;" "I *am*, therefore I perceive."

According to the atheistic/materialistic view of reality, how does a chemical, atom, or molecule suddenly step back and look at itself? Do the liver cells in my body wonder why they aren't kidney cells? Do the kidney cells wonder why they didn't end up as part of the optic nerve? Do they chalk it up to kismet, destiny, or karma? Molecules and chemicals don't have "identity." A human body that is only composed of material elements is nothing more than a machine, and a machine does not have self-awareness. It seems nonsensical to suggest such a thing and at the very least is profoundly counterintuitive. Even atheistic biochemist Dr. Stuart Kauffman chafes at such a notion:

> The second predominant view among cognitive scientists is that consciousness arises when enough computational elements are

> networked together. In this view, a mind is a machine, and a complex set of buckets of water pouring water into one another would become conscious, I just cannot believe this.[4]

That of course is the reason why Baggini and Dawkins declare that "most atheists...*believe*" that out of the exclusively physical "stuff" comes consciousness, emotion, morality, beauty, etc. For some reason they expect us to accept this idea on their authority. The obvious and most elegant solution to the "problem of consciousness," as Philosopher Daniel Dennet calls it (it's a "problem" for him because he *a priori* has made the decision that there cannot be a non-material soul), is that the "I" is spiritual.

The human body is essentially a sophisticated machine. For example, the heart is nothing more than a pump. The kidney is a filter, etc. Our bioengineering firms are able to produce replacement parts for the "human machine." Although relatively primitive, we have manufactured artificial limbs, knee joints, lungs, kidneys, and hearts. It is entirely conceivable that with advances in technology we will produce a pump that is superior to the human heart. The same is true for all of the aforementioned parts of the human body. However, it is the "I" that distinguishes the human being from the human "machine."

Not only is our technology totally incapable of building a machine that possesses a conscious "I", we shall soon see that science hasn't even a clue where to begin to attempt to accomplish such a feat. The only entity capable of projecting a conscious human "self" *is the "self" itself*! I would like to suggest to the reader a simple reason why the "I" cannot be duplicated by a machine: because the "I" is not an atom, molecule, or chemical; that the "I" exists in time but not in space; that the "I" constantly interacts with and affects the physical world (particularly the brain) but is itself undetectable and un-measurable by any material means of detection or measurement; that the "I" is the soul. Not only does this answer mesh with the deep, powerful intuitive feeling of most human beings, it is unmistakably the simplest, clearest, and easiest answer to the question. At the very least, there is no good reason why the honest skeptic should not seriously consider the possibility.

When Confronted with an Uncomfortable Idea, Deny, Deny, Deny

What we actually find, however, is that many skeptics and atheists are unable to approach the issue with open minds. They seem incapable of even *considering* such a possibility and simply go into denial. In a preposterously distorted argument, they claim that since no one can *physically locate* the "self" or the "soul," the logical conclusion must be that it does not really exist:

> **Despite our every instinct to the contrary**, there is one thing that consciousness is not; some deep entity inside the brain that corresponds to the "self," some kernel of awareness that runs the show, as the "man behind the curtain" manipulated the illusion of a powerful magician in *The Wizard of Oz*. After more than a century of looking for it, brain researchers have long since concluded that there is no conceivable place for such a self to be located in the **physical brain, and that it simply doesn't exist**.[5] (Journalist Michael Leminick, *Time Magazine*)
>
> There is no discrete self or ego lurking like a Minotaur in the **labyrinth of the brain**. There is no region of or stream of neural processing that occupies a privileged position with respect to our personhood... In subjective terms, however, there seems to be one... What are we conscious of?...**We are — we think — conscious of ourselves in our bodies**. After all, most of us don't feel merely identical to our bodies. We feel, most of the time, like we are riding around inside our bodies, as though we are an inner subject that can utilize the body as a kind of object. This last representation is an illusion...[6] (Dr. Sam Harris, atheistic neuroscientist)
>
> The intuitive feeling we have that there's an executive "I" that sits in a control room of our brain...is an illusion. Consciousness turns out to consist of a maelstrom of events distributed across the brain.[7] (Dr. Steven Pinker, Harvard University)

How, "despite our every instinct to the contrary," do we *know* that consciousness is not really a "self," that there is no "man behind the curtain" or "executive I" that runs the show, and that our perception of ourselves as being separate from our physical bodies is an illusion? Simple, declare these non-theists: since scientists have not found a particular *physical location* for the "self" inside the brain, then obviously there is no "self"!

The distorted thinking process that has actually transpired is the following: Instead of examining the data and evidence and then deciding if a spiritual/non-material dimension is plausible, the atheist declares *a priori* that there is no such thing, and proceeds to evaluate all data in light of that declaration. The deeply flawed syllogism goes something like this:

1. **(A)** Materialists propose as a premise, effectively as an article of faith, that there *cannot* be any spiritual, non-material dimension of existence:
   - "My fundamental premise about the brain is that its workings — what we call 'mind,' are a consequence of its anatomy and physiology and nothing more."[8] (Carl Sagan, astronomer)
   - "Everything, including that which happens in our brains, depends on these and only on these: a set of fixed, deterministic laws."[9] (Marvin Minsky, renowned expert in artificial intelligence)
   - "Resolutely shunning the supernatural...it *must* be in virtue of some natural property of the brain that organisms are conscious. There just *has* to be some explanation for how brains [interact] with minds... Consciousness, in short, *must* be a natural phenomena [*sic*]."[10] (Colin McGinn, philosopher)
   - "Our behavior is [solely] the product of physical processes in the brain."[11] (Steven Pinker, cognitive scientist)
2. **(B)** Therefore if the "self" exists, it *must* have a physical, material location inside the brain.
3. **(C)** Since we cannot find a specific physical location for the "self" inside the brain, the "self" does not actually exist and is an illusion created by the brain.

Sociobiologist and entomologist Edmund O. Wilson travels even further down this path of deeply flawed reasoning: "The brain and its satellite glands have now been probed to the point where no particular site remains that can reasonably be supposed to harbor a *non-physical soul.*"[12] Did Wilson actually presume to discover a non-physical soul with *physical* "probes"? Did he expect to find the soul by measuring electromagnetic waves? By definition, it is impossible to find something non-material by using material measurements; by definition, it is not something that can be seen, heard, touched, smelled, measured, or weighed. Its *effect* on our physical being can be seen and measured, which is why they were "probing" for it in the first place, but not the soul itself.

Perhaps it's time for the atheist to quit being so dogmatically narrow-minded and think a little outside the box. Perhaps the reason why every human being who has ever lived is so certain that their "self" is real is because it *is* real. Perhaps the obvious conclusion to draw from the fact that scientists have been unable to identify a physical location for the "self" is not that it is an illusion, but that the very real existent self in fact *has no physical location*; that it exists in time but not space, that the "self" is spiritual. How is it possible for an intelligent man like Wilson not to realize that the most compelling reason of all to seriously consider the possibility that the "self" is spiritual *is the very fact that we perceive and experience its effects so clearly and pervasively in our lives, while at the same time we are thoroughly incapable of measuring, defining, or quantifying it by any conceivable material method or standard?*

In fact, there is a simple method for the atheist/materialist to vanquish the theist and conclusively demonstrate the truth of his viewpoint. Nobody harbors any illusions that even the most sophisticated computer game on the market is anything other than the result of purely material processes. A skilled computer engineer is able to explain how all the high definition graphics — the soldiers, criminals, athletes, and other assorted characters running around in a computer game — are functions of algorithms and principles of electronics and physics. Keeping this in mind, it becomes obvious that the atheist has a very simple way to demonstrate that there is no soul or self that is separate from the physical brain.

All he has to do, just like the computer engineer, is to explain the chemical and molecular processes in the brain that produce consciousness, self-awareness, and the illusion of the executive "I" who runs the show. Alternatively, he could build a *machine* that has consciousness and self-awareness, thus demonstrating that physical processes alone are responsible. Could any assignment be simpler and clearer than that? As it turns out, it is one of those things much easier said than done. What insights does the world of science actually have to offer us on this subject? Consider the following:

- "**Nobody has the slightest idea how anything material could be conscious**. Nobody even knows what it would be like to have the slightest idea about how anything material could be conscious." [13] (Dr. Jerry Fodor, professor of philosophy and cognitive scientist, Rutgers University)
- "The problem of consciousness tends to embarrass biologists. Taking it to be an aspect of living things, they feel they should know about it and be able to tell physicists about it, **whereas they have nothing relevant to say**."[14] "**Consciousness seems to me to be wholly impervious to science**. It does not lie as an indigestible element within science, but just the opposite: science is the highly digestible element within consciousness..."[15] (Dr. George Wald, Nobel Prize-winning biologist)
- "The Hard Problem...is why there is first-person subjective experience. The Hard Problem is explaining how subjective experience arises from neural computation. The problem is hard because no one knows what a solution might look like...**everyone agrees that the Hard Problem remains a mystery**."[16] "The human brain is the most complex object in the known universe... No scientific problem compares to it... One challenge is that **we are still clueless about how the brain represents the content of our thoughts and feelings**."[17] (Dr. Steven Pinker)
- "What is consciousness? Well, I don't know how to define it. **I think this is not the moment to attempt to define consciousness, since we do not know what it is**."[18] (Sir Roger Penrose, mathematical physicist)
- "**Science's biggest mystery is the nature of consciousness**. It is not that we possess bad or imperfect theories of human awareness; we simply have no such theories at all. About all we know about con-

sciousness is that it has something to do with the head, rather than the foot."[19] (Dr. Nick Herbert, physicist)

If the phenomena of consciousness and self-awareness are such profound scientific mysteries, it would stand to reason that scientific opinions about the nature of the "self" (which is at the very center of consciousness and self-awareness), *if expressed at all*, would be offered in the most humble, guarded, and tenuous of terms. However, even after admitting that he has no meaningful understanding of the "self" and consciousness, that he has "nothing relevant to say" on the topic, and despite the fact that it is a subject which is "wholly impervious to science," the skeptic recklessly and obliviously drones on...

Speculation Masquerading as Science

Dr. Steven Pinker, cognitive scientist, informs us of the following:

> I don't believe there is such a thing as...a ghost in the machine, a spirit, or a soul, that somehow reads the TV screen of the senses and pushes buttons and pulls levers of behavior. There's no sense we can make of that...our behavior is the product of physical processes in the brain.[20]

"I don't *believe* there is...a ghost in the machine." Another true believer! The only things here that don't make sense are Pinker's incoherent statements about the nature of human behavior. If what Pinker says is true, then *who* analyzed the evidence and reached the conclusion that "our behavior is the product of physical processes in the brain?" *Who* decided that there is no soul and that the "self" is an illusion? *Who* is *fully aware and conscious* of the fact that there is no distinct "self" pushing buttons and pulling levers? If there is no *who* evaluating and deciding, all that is left is raw data and sensory information; in other words, what remains is a *what*, not a *who*. Information cannot understand, evaluate, or make decisions by itself *because it has no self*; it has no power, it has no ability to act, it has no faculty of will. This is self-evidently true whether the information and data in question are photographs from the Hubble Space Telescope, test results from a hospital laboratory, or sensory data and information stored in a human brain.

Data and information are totally inert and useless unless perceived, examined, evaluated, and acted upon by a *who*, i.e., a "self." To suggest that the brain automatically analyzes data and issues instructions like a sophisticated computer controlling a factory assembly line would be a flagrant begging of the question. The computer itself is useless without the assistance of a very real *who* in the form of a programmer or software engineer who gives the computer instructions. Pinker is very aware of the significance of this dilemma, but is determined at all costs not allow a separate "self" to skulk into the picture. When asked in an interview how we *know* that there is no "ghost in the machine", i.e., a spirit and soul that makes decisions, Pinker responded with the following:

> Well, we certainly can't prove it [translation: he really **doesn't** know] but on the other hand we have no reason to believe it either...the brain has a mind boggling complexity, 100 billion neurons, connected by 100 trillion synapses which is fully commensurate with the mind boggling complexity of the human mind. [Whose mind, Dr. Pinker, is being boggled?] If we opened up the skull and the brain was just spam or oatmeal, with no structure, we'd really have to think there is some magical extra ingredient. When you look at the staggering complexity of the neural network **you immediately see that it's capable of doing computations that we can't even dream of working out right now**, but the physical basis is there for complex intelligence.[21]

In other words, even though "nobody has the slightest idea how anything material could be conscious," even though scientists have "nothing relevant to say" on this topic, even though "everyone agrees that [consciousness] remains a mystery," *even though no scientist in the world has any idea whatsoever how the brain could generate consciousness or* ***if*** *it generates consciousness*, the non-oatmeal/spam brain has such incredible "mind boggling" complexity that you can "immediately see" that it's capable of accomplishing astounding feats. Thus, Pinker deduces that the brain is even capable of pulling off the fantastic trick of self-awareness, of fooling us into thinking that there is "someone" making decisions, and even creating the illusion that we are aware of ourselves making these decisions. (What a sparkling example of strictly scientific reasoning based on empirically established facts.) In short, what is Pinker's answer to the dilemma we posed above? *Abrakadabra, cerebral magic!* I don't begrudge Dr. Pinker the right to his belief in the magical, mystical abilities of neurons and synapses; that right is protected by the Constitution of the United States of America. However, I think he does display more than a little bit of chutzpah by trying to palm off his personal musings and predilections as science.

The root cause of this spiritual phobia on the part of the skeptic is wonderfully illustrated in this recounting by Ernest Becker in his Pulitzer Prize-winning book, *The Denial of Death*, of a conversation between Sigmund Freud and one of his closest disciples, Alfred Ernest Jones:

> Once while discussing psychic phenomena, Jones made the remark, "If one could believe in mental processes floating in the air, one could go on to a belief in angels," at which point Freud closed the discussion with the comment, "Quite so, even *der liebe Gott* [even the dear God].[22]

Once the possibility of a separate self is on the table, we are only a short step away from the possibility of God himself. In a January 2007 *Time* magazine article entitled "The Mystery of Consciousness," Dr. Pinker tells us that "no one knows what to do with the Hard Problem [how something material could be conscious]. Some people may see it as an opening to sneak the soul back in." Someone should inform the distinguished Dr. Pinker that we do not need to "sneak" the soul back in, because despite the unfounded and foolish assertions to the contrary, we never got rid of it in the first place.

If the skeptic is not prepared to accept or consider the idea of a spiritual soul that is separate from the brain, at least an unconditional acknowledgment of ignorance or lack of clarity on the matter would be a refreshing and admirable demonstration of intellectual integrity. Instead — much like a child who sticks his fingers in his ears while obnoxiously chanting, "I can't hear you" — the skeptic/atheist/materialist triumphantly declares that the "self" obviously *does not exist*! Actually, when

Pinker tells us that the self does not exist, he is experiencing an *illusion* that "he" is declaring or concluding anything; although we are then still stuck with the problem of *who* is experiencing the illusion.

Atheists Are Prepared to Deny Our Very Grasp on Reality

Atheists are prepared to burrow very deep down the materialist rabbit hole in order to avoid any possible confrontation with the spiritual. How deep? Deep enough to cast doubt on our very connection with reality. The skeptic claims that a scientific investigation of the brain leads us to the conclusion that there is no spirit or soul controlling our actions, and that the intuitive notion that there resides within us a separate "executive self" is an illusion. Leaving totally aside the issue of whether or not that assessment of the data is accurate, there is a much more fundamental question that must be addressed: *By what unique entitlement, privilege, or faculty does the skeptic confidently disavow as illusory the all-pervasive notion of a separate "self," yet simultaneously justify his absolute trust in his own perceptions and analysis regarding the "scientific" examination of the brain that led him to reach that conclusion in the first place?*

In other words, the crippling flaw of the atheist/materialist position is starkly highlighted by a simple fact: Our awareness of possessing a unique sense of consciousness (i.e., the "self") does not lie within our brain; *our awareness of possessing a brain lies within our unique sense of consciousness.* The "self" is not a projection of the brain. *Our knowledge of the brain is a projection or perception of the "self."*

Our entire interface with what we perceive and define as reality starts with our awareness of "self" and is processed through the "self." If that is an illusion, what is to stop us from concluding that everything we perceive is an illusion, including scientific research? By declaring the "self" as illusory, the materialist has breached the walls that protect our grip on a coherent sense of reality. Our entire concept of reality presupposes a distinction between the perceiver and the perceived. It is axiomatic that it is the "self" or the "executive I" who evaluates, discerns, and decides what is real and what is fantasy; once that is gone, our access to reality has been obliterated. All that remains is a series of perceptions and sensations whose true source and nature are unverifiable. There is no longer any distinction between "me" and what I perceive. There is no longer any difference between our "illusory" perception of the "self," our "illusory" perception that we possess a brain, and our "illusory" perceptions about a scientific examination of that brain. Everything is now up for grabs, including our ability to make decisions. Atheist academic Dr. Susan Blackmore informs us that:

> The "self" is not the initiator of actions, it does not "have" consciousness, and it does not "do" the deliberating. There is no truth to the idea of an inner "self" inside my body that controls the body and is conscious. Since this idea is false so is the idea of my conscious-self having free will...[it is a] false idea that there is someone inside who is in charge... Free will, like the "self" who "has" it, is an illusion.[23]

It may seem hard to believe, but there is certain logic behind this madness. Once you have accepted on faith that "our behavior is [solely] the product of physical processes of the brain," then it follows that not only is the "self" an illusion, but the intuitive feeling that the "self" is making choices, is also an illusion. Blackmore is in distinguished company. Nobel laureate Bertrand Russell seems to concur regarding the illusory nature of free will: "The first dogma which I came to disbelieve was that of free will. It seemed to me that all notions of matter were determined by the laws of dynamics and could not therefore be influenced by human wills."[24] This statement by Russell is not just nonsense of a *high* order; it is nonsense of a *cosmic* and *celestial* order! If there is no free will, not only would it be absurd to suggest that a person could "come to disbelieve" anything, it would be absurd to suggest that one could "come to *believe*" anything either. Everything is "determined" by the "laws of dynamics," like a series of billiard balls hitting each other. Just as the billiard ball has no control over its direction, velocity, or the force with which it collides with the next ball, a person has no control over what he or she believes or doesn't believe. If one believes in God, it is because an inescapable series of causes and effects determined that belief. If one is an atheist, it would be the result of exactly the same type of process.

When Russell believed in free will, it had nothing to do with his understanding or choices, and when he ceased believing in free will, it had nothing to do with his understanding or choices as well. It was *inevitable* that he would deny the dogma of free will. In fact, if we follow this line of reasoning to its logical conclusion, all discussion, argument, evidence, debate, and even scientific investigation on any subject would be completely useless and pointless. Everything we believe, perceive, understand, contemplate, all conclusions we reach, etc., are ineluctably ordained as the inevitable results of a series of causes and effects that are governed by the laws of physics and chemistry. After coming to the above conclusions, I made the pleasant discovery that on this point *I* am in distinguished company. The famed British philosopher of science Sir Karl Popper writes that:

> Physical determinism is a theory which, if it is true, is not arguable, since it must explain all our reactions, **including what appear to us as beliefs based on arguments, as due to purely physical conditions**. Purely physical conditions, including our physical environment, make us say or accept whatever we say or accept.[25]

According to the atheistic/materialistic view, we do not really think our own thoughts, experience our own lives, or decide our own actions. It is all just an inescapable series of deterministic events. If this sounds so outrageously absurd that one feels hard-pressed to believe that such an opinion could actually be expressed, think again. Here is atheist author Sam Harris:

> "While most of us go through life feeling like we are the thinker of our thoughts and the experiencer of our experiences, **from the perspective of science we know that this is a false view**."[26]

I wonder...are we the *speakers of our speeches and the writers of our writings*? In this profoundly confused worldview, the "self" is an illusion, free will is an illusion, we don't think our own thoughts, we don't experience our own experiences, and in fact, outside of our own thoughts, there is no verifiable reality at all. All of this provides an illuminating backdrop for the following episode that is described by Dr. David Berlinski in his classic work, *The Devil's Delusion: Atheism and Its Scientific Pretensions*:

> In a recent BBC program entitled "A Brief History of Unbelief," the host, Jonathan Miller, and his guest, philosopher Colin McGinn, engaged in a veritable orgy of competitive skepticism, so much so that in the end, the viewer was left wondering whether either man believed sincerely in the existence of the other.[27]

To be fair, not every atheist flatly denies the existence of free will as do Bertrand Russell, Susan Blackmore, and Cornell University's Will Provine. Prominent atheist/materialists such as Daniel Dennett and Richard Dawkins desperately struggle to straddle both sides of the fence and retain some concept of free will (unsuccessfully, as far as I'm concerned), while at the same time carefully trying avoid the "trap" of admitting the existence of a "self" (which is exercising that will) that is separate from the brain. I would also add that even those non-theists who express these ridiculous ideas don't actually live their lives as if they are true.* From their pulpits in the rarified world of the university lecture hall, they declare the "self," free will, thoughts, and experience as illusory, and then go back to everyday living like the rest of mankind. In other words, they continue living and acting as if their "self" is absolutely real, as if they have free will, and as if their own thoughts and experiences truly belong to them.

What *is* the so-called "scientific" basis for a strictly materialistic view of the brain and consciousness? Dr. Steven Pinker briefly summarizes his approach in an article entitled "The Mystery of Consciousness" that appeared in *Time* magazine in January 2007:

> Scientists have exorcised the ghost from the machine not because they are mechanistic killjoys, but because they have amassed evidence that every aspect of consciousness can be tied to the brain. Using functional MRI, cognitive neuroscientists can almost read people's thoughts from the blood flow in their brains. They can

* Imagine Susan Blackmore buying paint for her living room. The clerk asks her if she would like the peach color or the off-white. She answers, "I *must choose the peach!*"

> tell, for instance, whether a person is thinking about a face or a place or whether the picture the person is looking at is of a bottle or a shoe... And consciousness can be pushed around by physical manipulations. Electrical stimulation of the brain during surgery can cause a person to have hallucinations that are indistinguishable from reality, such as a song playing in the room or a childhood birthday party. Chemicals that affect the brain, from caffeine and alcohol to Prozac and LSD, can profoundly alter how people think, feel and see.[28]

It's very difficult to understand what relevance any of the above has to the question of the existence or non-existence of a separate self or a soul. Let's go through it point by point.

- I decide to have a conversation with a friend. A cognitive neuroscientist in another room monitoring the brain activity correctly assesses that I am involved in a conversation. I then decide to end the conversation and begin contemplating a mathematical problem that has been bothering me for a long time. The neuroscientist again correctly assesses that I am now involved in abstract reasoning. How does that prove that I don't have a soul or make free-will decisions? While there is no doubt that the brain is *involved* in the process of both the conversation and my contemplation of mathematics, it is also clear that more is going on than just the sum total of neural and cerebral activity. It was "my" (i.e., my soul's) decision to have a conversation that caused a particular part of my brain to light up; it was "my" decision to think about mathematics that caused a different part of my brain to light up. This clearly points to a separate self who is in control. The crucial issue is not the fact that scientists can identify the different parts of the brain that are involved in different functions, but what is the connection between my decision and my brain? How does my brain know to switch cerebral circuits when I decide to switch my activity? How and why does my will turn on and off different parts of the brain? How does "my" brain convey information to "me"?
- A specific part of my brain is electrically stimulated during surgery and I vividly relive the experience of my mother singing a lullaby to

me when I was a child. Again I ask, what does this have to do with the existence of the soul, and again Pinker has ignored the essential point. The issue that needs to be addressed is not ***that*** I experience the song when the brain is poked in the right way, but "who" is experiencing the song right now, "who" experienced the song in childhood, and how is it possible for anything material to be conscious and self-aware of experiences like lullabies to begin with?

- *"Chemicals that affect the brain...can profoundly alter how people think, feel, and see."* What is Pinker's point here — that artificial manipulation of the brain affects me? Was that actually ever in question? If getting punched in the stomach affects my brain and can alter how I think, feel, and see, why would I expect a lesser result if the brain itself is artificially manipulated? Is he trying to tell us that before the advent of modern neuroscience nobody ever noticed that alcohol affects people's behavior? A slight rephrasing of Pinker's words highlights the truly relevant issues that are being ignored: "Chemicals that affect the brain...can profoundly alter how 'I' think, feel, and see." *Who* is thinking, *who* is feeling, *who* is seeing (and *who* is getting drunk and stoned?) and how is it possible for *me* to think, feel, and see in the first place?

How is it possible for Pinker to state on the one hand that it is a total "mystery" how "first-person subjective experience...arises from neural computation," that "we are still clueless about how the brain represents the content of our thoughts and feelings," and then a moment later state that scientists have "amassed evidence that *every aspect* of consciousness can be tied to the brain"? Yes Dr. Pinker, *every* aspect of consciousness, except *what it is, how it works, where it comes from, and how such a thing could be possible in the first place*! The evidence that he presents only tells us that the physical brain is *involved* in many aspects of consciousness (which frankly is not the type of revelation that is likely to launch a thousand ships); it does not even begin to tell us what consciousness is and it does not in any way contradict the notion of a separate self or soul that works in tandem with the brain. In fact, all the "scientific evidence" that Pinker presents is nothing more than an elaborate repetition of what physicist Nick Herbert had already stated so simply and elegantly: "About all we know about consciousness is that it has something to do with the head and not the foot."

I once debated the well-known atheist polemicist, Dr. Michael Shermer, on the *Michael Medved Radio Show*. He claimed that diseases like Alzheimer's are compelling evidence that there is no soul. As the brain deteriorates, we observe that the personality also deteriorates, so based on this, Shermer concluded that personality and the self are identical with the physical brain. I assume Pinker would draw the same conclusion. This is a deeply flawed argument and in fact misses the mark entirely. The point of contention between theist and non-theist is not about the fundamental importance of the astounding piece of organic machinery called the human brain. It is whether or not there is a non-material soul/mind that works *together* with the brain. If a person is unable to speak after a stroke, it does not mean that his personality has disappeared; it means that *our accessibility* to this person's personality has been limited by the damage to the speech centers of the brain. If they are unable to write, it means *our accessibility* to their personality has been limited even more by the damage to the brain. It is obvious to all that with extensive damage to the brain, whether by disease or by death from a blow to the head with a lead pipe, our *access to another individual's personality* disappears altogether. Nobody knows what actually happens to the soul/personality in such cases. The assertion that because we cannot access another's personality due to brain damage, the self is wholly a function of brain activity is — ironically enough — a stunning example of an Argument from Ignorance:

- Shermer and Pinker, along with every other scientist on the face of the Earth, haven't the slightest idea how *or if* the brain generates consciousness, self-awareness, and personality.
- Since they are completely ignorant of the nature of the fundamental link between the brain and the self, they are by definition completely ignorant of the nature of the fundamental link between brain damage and the subsequent limited accessibility to the personality of the afflicted person. In order to understand one, you must understand the

other. If one doesn't understand how consciousness is generated and what consciousness is in the first place, it is absurd to claim you know where it went when you cannot see it anymore.

- Simply put: Shermer and Pinker *don't know* what the conscious self actually is when the brain is functioning normally, and they *don't know* what actually happens to the conscious self when the brain is damaged. Hence, any definitive statement about the intrinsic nature of consciousness based on the phenomena of brain damage is an argument based on *ignorance of the highest order.*

Mind over Matter: "I" Can Change My Brain

As we have discussed, the atheist claims that the "self" who is in control of decision-making is an illusion created by the brain. If that were true, then it would certainly be impossible for me to physically affect my brain through my own willful decisions. How could an illusion created by the brain actually turn around and alter brain function in a way that could be scientifically documented? It would be as if a character in a computer game actually turned around and changed the program, function, and hardware of the game itself; a wonderful idea for a science fiction story but impossible in the real world. Evidence of such phenomena would be a crushing blow to the materialist view of reality.

Via the latest brain imaging technology, Canadian neuroscientist Dr. Mario Beauregard, among many others, has documented that "I" can actually change the patterns of activity in my brain through the exercise of my will. In the following passage he briefly summarizes some of his findings:

> Materialists must believe that their minds are simply an illusion created by the workings of their brain and therefore that free will does not really exist and could have no influence in controlling any disorder. But nonmaterialist approaches have clearly demonstrated mental health benefits... Jeffrey Schwartz...a UCLA neuropsychiatrist, treats obsessive-compulsive disorder — a neuropsychiatric disease marked by distressing, intrusive, and unwanted thoughts — by getting patients to reprogram their brains. Their

minds change their brains. Similarly, some of my neuroscientist colleagues at the University of Montreal and I have demonstrated, via brain imaging techniques, the following:

- *Women and young girls can voluntarily control their level of response to sad thoughts, though young girls found it more difficult to do so.*
- *Men who view erotic films are quite able to control their responses to them when asked to do so.*
- *People who suffer from phobias such as spider phobia can reorganize their brains so that they lose the fear.*
- *Evidence of the mind's control over the brain is actually captured in these studies. There* **is** *such a thing as "mind over matter." We do have will power, consciousness, and emotions, and combined with a sense of purpose and meaning, we can effect change.*[29]

Dr. Reuven Feuerstein was an educational and cognitive psychologist who studied under one of the great pioneers in cognitive psychology, Dr. Jean Piaget. Feuerstein was the founder and director of the International Center for the Enhancement of Learning Potential (ICELP) in Jerusalem. The staff and faculty of ICELP, putting into practice the protocols and techniques developed by Feuerstein, have achieved astounding results in restoring function and dramatically increasing learning potential in people with conditions such as: autism, Down syndrome, and those who have suffered brain damage in terrorist attacks, strokes, drownings and car accidents. In November of 2008, Feuerstein addressed the National Urban Alliance:

> Human beings are modifiable and not just modifiable in terms of their behavior...not just the structure of their behavior can be modified, but actually the neural system can be modified, and marvelously, miraculously. We can point out the fact that the behavior which we impose on our brain, modifies our brain. It means it's not only the brain which shapes our behavior, **our behavior shapes the nature of the hardware of our brain**.[30]

As the aforementioned Dr. Jeffrey Schwartz so appropriately concludes: "The time has come for [those who advocate a purely materialistic view of] science to confront the serious implications of the fact that directed, willed mental activity can clearly and systematically alter brain function." [31]

The actuality of a "self" that is separate from our physical being and can change the physical configuration of our neural hardware is not the end point of our exploration of spirituality — it is just the beginning. The spiritual ocean we swim in is quite large...

Soul Power: Communication through Speech

What is actually happening when you have a conversation with another person? We do it so often and naturally that it is taken for granted. If you have children, then you have watched your children learn how to talk. Did they think thoughts before they learned how to talk? In what language were they thinking? How does a young child conceive the thought that he wants a drink when he does not know any words at all? After all, children who do not know how to speak a language are still able to communicate (and they still get thirsty). They can make it very clear to their parents or caregivers that they have wants and needs. My three-year-old daughter many times would translate for us when we could not understand what her one-year old brother wanted. He would be making sounds that to us were incomprehensible and somehow she could tell he wanted a particular toy, or a drink, etc. The critical question however, is what was going on in *his* head as he was speaking this incomprehensible language? No matter what explanation you may offer, it's clear he was thinking ideas that he did not yet know how to translate into language.

"Words" and "Ideas" Are Two Separate Things

The *thought* "I want a drink" is not the same thing as the *words* "I want a drink." In fact, the two in a certain sense have absolutely nothing to do with each other. My daughter understood by certain sounds her little brother made that he was thinking inside his head "I want a drink," even though he could not yet speak the words and in fact he was even too young to *think* the words. *The words themselves mean nothing at all.* They are simply arbitrary sounds that we use to express an idea. The obvious proof of this is that if someone says "I want a drink" in a foreign language, we have no idea what they are talking about.

I have a thought in my head: "I desire to imbibe liquid into my body." I then proceed to form a series of *arbitrary, intrinsically senseless* sounds with my mouth. These sounds travel through the air where they are heard by another person, who then decodes them and understands to bring me a glass of water. We take it for granted because we do it all the time, but what is transpiring is nothing short of miraculous. I am taking *ideas* in my head and sending them through the air to other people. *I am attaching ideas to sound waves.* Remember, the sound or word is not the same as the idea, concept, or the particular information I am communicating.

Imagine I am in a room with a man who only speaks Chinese and it becomes clear to him that I have no idea what he is trying to tell me. He then points to his throat and makes a gesture with his hand as if he is drinking something. I suddenly realize he is thirsty and wants a drink. That is to say: we are now thinking and sharing the exact same *idea*, but we convey it using different *sounds*. The word is an arbitrary, meaningless sound, but the information, idea, or message is very specific and very meaningful. They are two completely unrelated things and yet somehow I am able to bind them together. Understanding language means understanding *ideas and information*, not sounds.

Let's try to understand this from the perspective of the materialist scientist. What parts of a conversation will he be able to physically measure and quantify in the laboratory? He can monitor the brain activity from the person speaking, he can measure the sound waves as they travel through the air, and he can monitor brain activity in the person listening to the words. Can the scientist measure, quantify, or define the "idea," the "information," or "message" that passes between two people? Can he "touch" the idea? Can he "smell" the idea? Can he "see" the idea? Can he "taste" the idea? Can he "weigh" the idea? In truth he cannot even "hear" the information. He can only understand what is being said if he also speaks the language, but that is just begging the question. The idea itself, "I want a drink of water," is not accessible through any of our physical senses, nor can it be quantified or measured for the simple reason that it is not physical or material in any way. If I speak complete gibberish in the same tone and for the same amount of time, the measurements of the scientist will be the same: electrical activity and sound waves. Evolutionary biologist George Williams put it this way: "You can speak of galaxies and particles of dust in the same terms because they both have mass and charge and length and width. [But] you can't do that with *information* and matter... Information doesn't have mass or charge or length in millimeters..."[32]

Even Dr. Steven Pinker, the highly celebrated Harvard professor specializing in cognition and linguistics, can barely contain his wonder and awe when contemplating the human capacity for language. This self-declared "proud atheist" manages to seamlessly weave the "m" word into his description of that remarkable faculty called language in the very first paragraph of his award-winning book, *The Language Instinct: How the Mind Creates Language*:

> "You are taking part in one of the wonders of the natural world. For you and I are members of a species with a remarkable ability... That ability is language... The ability comes so naturally that we are apt to forget what a **miracle** it is."[33]

A miracle, indeed! All day long we are involved in binding together the spiritual reality of ideas and thoughts with the physical reality of sound. Speech is nothing less than one soul relaying a spiritual message to another soul through the physical medium of sound. We are just so used to it that we never take the trouble to think about what is actually happening.

The Miracle of Written Language

Many years ago, I saw a film about a Catholic missionary who was sent to the Native Americans living in the Great Lakes area of the North American Continent.[34] The story takes place in the early part of the seventeenth century. These Indians had no written language. In one scene they express their bewilderment as to what the white men are doing scratching strange symbols on pieces of paper. The priest finds that he cannot explain it to them conceptually, so he asks one of the Indians to say a word. He says "canoe." The priest writes the word and then he and the Indians who are watching walk ten or fifteen yards over to another of the white men. The priest then asks this man to read the word. He reads "canoe." The Indians jump back in horrified astonishment. They start screaming at the priest that what he just did could only be the work of a demon.

Of course we just shake our heads and condescendingly smile at these primitives who would mistake something as simple and as elementary as written words for witchcraft or the work of a demon. However, the reason we do not share their horror and awe is not because the written word is so simple and elementary, it is only because we have become so used to the mind-boggling, fantastic notion of written language.

I have a thought in my head. I have an emotion that I am feeling. I have ideas I want to communicate. I want someone across the ocean to know how I feel, to know what I am thinking, to understand what I understand. How can I possibly accomplish such a feat? Very simply, I take a pen or pencil, scratch a few lines on a piece of paper, and someone thousands of miles away then can know my thoughts and feelings. If I scratch the right symbols I can bring a person to tears, to feel joy and happiness, to feel anger or disappointment. Not only can I attach non-material ideas and feelings to sound waves, I can attach them to lines on a piece of paper, or for that matter to symbols on a computer screen. I again bind the spiritual and physical.

Just like the sounds of a spoken language, letters mean *absolutely nothing*. They are random, arbitrary markings made with a pen, pencil, crayon, or keyboard. An English speaker looking at Arabic or Chinese writing for the first time would have no idea if he was looking at an alphabet or the random doodling of a child. Renowned Hungarian scientist and philosopher Michael Polanyi put it this way: "A book transmits information. But the transmission of the information cannot be represented in terms of chemical and physical principles. In other words, the operation of the book is not reducible to chemical terms."[35] Imagine the most sophisticated and advanced laboratory that exists in the world today, and have the scientists there analyze the paper and the markings on the paper. They can tell you the chemical composition and the molecular structure of the paper and the ink, but they cannot ever hope to measure or quantify the ideas and feelings that are contained in the words — *and yet they are there nonetheless*. The only monitor capable of detecting the ideas, concepts, and emotions contained in the sound of a human voice or in the writing of a human poet is the "monitor" called the human soul.

Sending Messages via Light Waves

Not only can I bind a non-material/spiritual idea or piece of information to markings on a piece of paper and FedEx it to you overnight (or instantaneously by email); not only can I attach emotions, concepts, and information to a sound wave and send it to you at the speed of sound; I even have the ability to attach non-material messages to a *light wave* and send it to you at the speed of light without the aid of a video camera, satellite, telephone, or computer. Does it sound too fantastic to believe, like science fiction? It is just another illustration of how routine and habit can dull our awareness to the miraculous things going on around us all the time.

A loved one is in surgery. You are anxiously awaiting word on the outcome of the delicate and critical procedure. At the other end of the waiting room you see the chief surgeon emerge still dressed in his scrubs. He catches your eye, flashes a big wide smile, gives you the thumbs up, and then disappears back behind the swinging doors. You breathe a sigh of relief and whisper a prayer of thanks (if you're not an atheist). What just happened?

If I were to offer a scientific description of this event it would go something like this: Light waves traveling at approximately 186,000 miles per second were reflected off the figure in surgical scrubs, were perceived by your eye, and via your optic nerve were sent to your brain where the image was analyzed. Your brain told you that you were looking at a surgeon with his lips pulled back exposing his teeth and his thumb extended upward. You concluded that the doctor was conveying to you a message that the surgery was successful and the patient will be fine. How was the doctor able to send you the message and how were you able to understand the message?

As strange as it may sound, what happened was that the doctor attached a spiritual entity (information) to a light wave and sent it to you. His physical gestures were the language, but it was a language that arrived via light. The information was *not* "somebody's lips pulled back exposing their teeth" (a smile), and the information was *not* "an extended thumb" (thumbs up). The information was that *the surgery was successful and the patient is doing fine.* One of the powers of the human soul is to take spiritual ideas and information, and bring them into the material world by attaching them to a physical gesture.

All thoughts, ideas, and information are spiritual entities that can only be brought into our material word by being attached to a material entity (i.e., writing, sound, gestures, etc.). Imagine someone under a type of anesthesia where he is fully conscious but cannot move or speak. Despite the fact that this person's head is filled with all kinds of thoughts, information, and feelings, no one else can possibly be aware of what they are. This simple reason is because the tools needed to bring these spiritual entities into our physical universe are inoperable. I repeat: the thoughts, ideas, feelings, and information exist, but they are "trapped" in the spiritual realm.

Self-Esteem: A Projection of Our Spiritual Essence

Most of us have skeletons in our closets. It may be something very big, or perhaps relatively small, but just about everyone has done things that they hope and pray no one ever finds out about. Picture in your mind that particular "skeleton" and imagine being at a joyous public gathering of those who are close to you (birthday party, engagement party, wedding, etc.), and someone reveals to everyone assembled the exact details of that terrible thing that you did — the deed that you had hoped would stay secret for all eternity. What is your reaction? Most likely an indescribably painful feeling of humiliation. If you were hooked up to an array of body sensors, they would certainly record a sudden, profound change in your body chemistry, heart rate, brain waves, etc.

What exactly happened that caused your physical bodily functions to go haywire? Did someone inject a chemical into your body? Did you inhale a gas? Did you ingest a solid or liquid? *Was there any change at all in the physical universe*? The only thing that changed was that something you had done, which up until now was a secret, became public knowledge. Is there a scientific measurement for quantifying when a secret is not a secret anymore? For that matter, is there a scientific measurement for a secret altogether? Is there any possible way to scientifically detect such a thing? How is it possible to destroy someone's self-esteem with cruel words and insults? Are there tiny pieces of chemicals that enter into the person along with the offending words? How is it possible to lift somebody with a genuine compliment or kind words? Are there mysterious molecular structures that float through the air that enter a person's body along with the heartfelt praise and acknowledgment? I defy any scientist to demonstrate that there is a material, chemical, or molecular basis to explain why vicious or compassionate words, scrawlings on a piece of paper, or physical gestures can have such a profound effect on another human being.

When Julian Baggini depicts the non-material world of the soul as an "implausible tale," he is grossly mistaken. Not only is it not an "implausible tale," but it is palpably and unmistakably an integral part of the reality in which we exist, as evidenced by our sense of self-awareness, the ability of the "self" to reprogram brain patterns, our relentless quest for transcendent meaning, the spoken word, written language, non-verbal communication, and our sense of self-esteem. It is also unquestionably true that the spiritual and physical interact and affect one another.

It is the non-theist who must demonstrate that the flow of ideas and information between human beings — whether in the form of the written word, spoken word, or physical gestures — can be accounted for and understood within the strict parameters of the laws of chemistry and physics. When I reveal an unpleasant secret about another human being, the atheist must demonstrate that a "secret" has material, chemical, and molecular properties that cause the physical and mental reactions we call "humiliation." It is the atheist who must demonstrate that consciousness, the self, ideas, thoughts, concepts, emotions, communication, and self-esteem are definable wholly in terms of physical "stuff." It is the materialist who must produce a plausible, empirically demonstrable alternative to the self-evident reality of the ocean of spirituality in which all mankind is immersed.

We have thus far seen Richard Dawkins and his colleagues attempting to escape the inescapable drive for transcendent meaning by chasing comforting fictions. We have also analyzed the failed attempts by Julian Baggini, Steven Pinker, and other skeptics to pigeonhole all human experience into the world of materialism. We will now investigate the last, and in a certain sense the most crucial, of the three major concepts that were mentioned at the beginning of Chapter 6: morality. I will attempt to demonstrate to the reader that true, absolute moral principles can only exist in a Godly universe, where human beings are inherently different than all other creatures. In the world of the non-theist there is neither true morality nor true immorality, only amorality...

End Notes

1 Baggini, *Atheism: A Very Short Introduction*, p. 30.

2 Ibid., p. 6–7.

3 Dawkins, *The God Delusion*, p. 34.

4 Dr. Stuart Kauffman, *Beyond Reductionism: Reinventing the Sacred*, http://www.edge.org/3rd_culture/kauffman06/kauffman06_index.html

5 Mario Beauregard and Denyse O'Leary, *The Spiritual Brain: A Neuroscientist's Case for the Existence of the Soul*, (New York: HarperCollins, 2007), p. 115.

6 Sam Harris, "Consciousness Without Faith," http://richarddawkins.net/articleComments,497,Consciousness-Without-Faith,Sam-Harris-On-Faith,page1#16807

7 Steven Pinker, "The Mystery of Consciousness," *Time Magazine*, January 19. 2007, http://www.time.com/time/magazine/article/0,9171,1580394-2,00.html (sec. 2)

8 Beauregard and O'Leary, *The Spiritual Brain*, p. 125.

9 Ibid., p. 30.

10 Colin McGinn, *The Problem of Consciousness: Essays Towards a Resolution* (Cambridge, Massachusetts: Blackwell Publishers, 1993), p. 6.

11 "Steven Pinker on the Ghost in the Machine, Soul, Free Will," http://www.youtube.com/watch?v=4A_r6_GGv3U

12 Beauregard and O'Leary, *The Spiritual Brain*, p. 3.

13 Dr. Jerry Fodor, "The Big Idea: Can There Be a Science of Mind?" *Times Literary Supplement* (July 3, 1992), p. 5.

14 George Wald, "The Cosmology of Life and Mind," in W. Harman with J. Clark, J., eds. *New*

Metaphysical Foundations of Modern Science, Institute of Noetic Sciences (1994), pp. 123–131.

15 George Wald, "Life and Mind in the Universe," *International Journal of Quantum Chemistry* (April 16, 2008).

16 Steven Pinker, "The Mystery of Consciousness."

17 Steven Pinker, "Will the Mind Figure Out How the Brain Works?" *Time Magazine*, April 10, 2000, http://www.time.com/time/magazine/article/0,9171,996604,00.html

18 Sir Roger Penrose, *The Large, the Small, and the Human Mind* (Cambridge University Press, 1997), p. 98.

19 Beauregard and O'Leary, *The Spiritual Brain*, p. 108.

20 "Steven Pinker on the Ghost in the Machine, Soul, Free Will," YouTube.com

21 Ibid.

22 Ernest Becker, *The Denial of Death* (New York, NY: Free Press Paperback, Simon & Schuster, Inc., 1997), p. 105.

23 Susan Blackmore, *The Meme Machine* (Oxford University Press, 1999), pp. 236–237.

24 Beauregard and O'Leary, *The Spiritual Brain* p. 30.

25 Karl Popper, *Objective Knowledge: An Evolutionary Approach*, (Oxford University Press, 1972), p. 223.

26 Harris, "Consciousness Without Faith."

27 David Berlinski, *The Devil's Delusion: Atheism and Its Scientific Pretensions* (New York: Crown Publishers, Random House, 2008), pp. 8–9.

28 Pinker, "The Mystery of Consciousness."

29 Beauregard and O'Leary, *The Spiritual Brain, xiii.*

30 Dr. Reuven Feuerstein addressing the Urban Alliance, http://www.youtube.com/user/NatUrbanAlliance#p/u/4/uXopVpQwivY

31 Beauregard and O'Leary, *The Spiritual Brain*, p. 126.

32 As cited by Dr. Stephen Meyer in *Signature in the Cell: DNA and the Evidence for Intelligent Design* (New York: HarperCollins, 2009), p. 15.

33 Steven Pinker, *The Language Instinct: How the Mind Creates Language* (New York: HarperCollins, 1995), p. 1.

34 *Black Robe,* a film directed by Bruce Beresford, 1991.

35 Michael Polanyi, "Life Transcending Physics and Chemistry," *Chemical and Engineering News*, Vol. 45, No. 35 (August 21, 1967).

Section 4

Atheism and Morality

Chapter 8

Amorality and Atheistic Values

"Amoral": (a) neither moral nor immoral; (b) outside the bounds of that to which moral judgments apply (Webster's Third New International Dictionary)

One inescapable implication of atheism is this: We are qualitatively no different than any other animal. George Gaylord Simpson, paleontologist, influential evolutionary biologist, and professor of Zoology at Columbia University, put it this way:

> Man is the result of a purposeless and materialistic process that did not have a human in mind. He was not planned. He is a state of matter, a form of life, a sort of animal, and a species in the order of primates, akin nearly or remotely, to all life, and indeed, to all that is material.[1]

Professor Brad DeLong of UC Berkeley, although perhaps lacking the eloquence of Simpson, makes the same point: "I'm just a jumped-up monkey."[2]

The world of the animal is *amoral*. It is "outside the bounds of that to which moral judgments apply." When a lion kills a zebra, the lion has not done anything "moral nor immoral." That is simply the way nature operates, the way things are. If it is true that we are part of the animal world, there is no difference between a lion killing a zebra, a person killing a zebra, and a person killing a person. Why is one viewed as the way nature operates and one suddenly labeled immoral? Morality, in a non-theistic context, is nothing more than a *word* that an individual or society assigns to behavior that it subjectively has decided it prefers. Subjective morality has no inherent significance. In other words, the fact that the zebra quite clearly prefers, subjectively, not to be eaten by the lion in no way at all changes the amoral nature of the act. In the same way, if a particular "human animal" would prefer not to be harmed by another "human animal," it does not magically create a mysterious entity called morality. In atheistic philosophy, "morality" means whatever the individual or the particular society wants it to mean.

> "Morality is the custom of one's country and the current feelings of one's peers. Cannibalism is moral in a cannibalistic country."[3]
> (Samuel Butler)

It follows logically that just as cannibalism is moral in a cannibalistic country, so too slavery is moral in a country that has slaves, lynching is moral in a society that accepts lynching, forced abortions are moral in a country that accepts forced abortions, and killing unwanted female infants is moral in a society where females are considered inferior and undesirable. It can be quite a shock to the system of the born-again secularist to realize that the cannibal in the Fiji Islands viewed a public feast consisting of a dozen cooked human beings with the same unremarkable, mundane attitude that we view a barbeque on the Fourth of July. He would react to our disgust at his practices with the same bewilderment that we would feel if someone protested our consumption of a hot dog. Who is to say that he is wrong?

Each society establishes, maintains, and modifies its values to suit its own needs. It is absurd to suggest that any particular individual or society has the authority to dictate to all human beings what their values should or shouldn't be. It is even more absurd to suggest that the pronouncements of a particular individual or society in any way *obligate* others to behave accordingly. An atheistic philosopher can say that a certain system of behavior does not appeal to his taste or that he even finds it abhorrent, but to call it *immoral* distorts the true nature of things. In the words of one of the most prominent atheistic philosophers of the twentieth century, Bertrand Russell:

> I cannot see how to refute the arguments for the subjectivity of ethical values, but I find myself incapable of believing that all that is wrong with wanton cruelty is that I don't like it.[4]

Russell's incredulousness notwithstanding, there is no escaping the truth of his own stated conclusion. For Russell the only thing that could possibly be wrong with cannibalism or torturing and molesting children is that he subjectively doesn't like it. In fact, a prominent twenty-first century atheist philosopher, Dr. Joel Marks, confirms this proposition:

> Even though words like "sinful" and "evil" come naturally to the tongue as a description of, say, child-molesting, they do not describe any actual properties of anything... There are no literal sins in the world because there is no literal God...I now maintain nothing is literally right or wrong because there is no Morality... Yet we human beings can still discover plenty of completely-naturally-explainable internal resources for motivating certain *preferences*. Thus, enough of us are sufficiently averse to the molesting of children, and would likely continue to be so.[5]

We see that in the crucible of an honest atheistic worldview, child molesting undergoes a metamorphosis from *sinful and evil* to being a matter of *preference*. Of course when it comes to preferences, some prefer this way and some prefer that way. It has been my experience that people often get confused when discussing this topic. Therefore, in an attempt to eliminate confusion, please read the following points carefully before continuing:

- Atheists *certainly* have values and principles that guide their lives and decisions. The word that one chooses to describe or conceptualize these values — be it *morality, ethics, utilitarianism, humanism, etc.* — is beside the point; the values are whatever they are, no matter what one calls them.

- There is no question about the *existence* of atheistic values; it is the *significance* of these values that is the crucial issue under discussion.
- My assumption is that the overwhelming majority of atheists who live in the United States of America and Canada strongly oppose the acceptance of pedophilia and child molestation. After all, they have grown up in a society that has a deep abhorrence to such practices.
- I am not accusing any atheistic philosopher of "ethics" of *approving* of pedophilia. I accuse them of *laying the philosophical groundwork* that opens the way for the acceptance and approval of pedophilia.
- Here is the central point: An honest and consistent articulation of an atheistic worldview must admit that "ethical" values have no significance at all outside of the heads of those who espouse them. They lack objective reality and any *significance* ascribed to these values is rooted squarely in human imagination and psychological conditioning. They are attempts to create the illusion that human actions and decisions have real purpose and meaning. In other words, their significance is as illusory as the significance of (what the non-believer would consider to be) my imaginary notion that God spoke to the Israelites at Mt. Sinai.
- Human beings have an innate sense of compassion, empathy, the need to bond and form relationships, and the ability to love. Human beings also have an innate sense of lust, selfishness and self-indulgence, the ability and desire to hate and to dominate, and experience powerful cravings and urges to act with brutality and cruelty. All of these are *natural* tendencies. In fact, it is self-apparent that in a purely material world, every type of behavior is *natural*. There isn't anything else. How an individual views these different emotions and drives and chooses to prioritize them are matters of personal preference. If one so desires, he can label these personal preferences with the words "moral" or "immoral"; the *word* that one chooses does not change the fact that they are nothing more than personal preference or societal conditioning. To put it a different way: in an atheistic world, the terms *morality* and *personal/societal preference* are identical and interchangeable.
- It is rather obvious that once one achieves an intellectual clarity that objections to pedophilia or any other type of behavior are nothing

more than *personal preferences* or are the result of *societal conditioning*, the door is wide open to consider other types of preferences and societal mores. Societies change and personal preferences change.

These principles are so basic that they could be labeled Moral Philosophy 101. They are understood and agreed upon by almost all serious thinkers, theistic and non-theistic. There is very little difficulty involved in the *intellectual* acknowledgement and acceptance of the truth of these ideas. However, there is enormous emotional and psychological resistance. For those raised in a society where the moral underpinnings are based on an absolute, unshakeable God-given Judeo-Christian tradition, it can be quite traumatic to realize that with the rejection of the God who is *the* moral authority, one is left with *no* moral authority at all — other than the shifting sands of public opinion and personal preference.

What continually puzzles me is that even after accepting that in a Godless world morality is nothing more than elegant window dressing to describe the latest public opinion poll, these philosophers continue to expound and search for some meaningful expression of moral principles. I have always had great difficulty detecting an intelligible, orderly flow in the logic when non-theistic thinkers present these expositions. A few examples....

Dr. Julian Baggini and Christopher Hitchens

In *Atheism: A Very Short Introduction*, Dr. Julian Baggini explains that "...religion is a human construct that does not correspond to any metaphysical reality." [6] Let's assume for argument's sake that what Dr. Baggini has written is true. How then would we view religion? Baggini's obvious point is that if religion is just a human construct with no connection to a higher reality, why should anyone take it seriously? It is just another product of the creative human imagination and has no intrinsic value or meaning at all. Baggini then proceeds to tell us:

> However, many people think that atheists believe there is no God and no morality... [However], there is nothing to stop atheists from believing in morality, a meaning for life, or human goodness.[7]

If Baggini's ears would only listen to what his own mouth is saying! Not only are *all* atheistic systems nothing more than "human constructs" that do not correspond to any "metaphysical (beyond the physical) reality," the atheist explicitly denies there *is* such a thing as a metaphysical reality. The only things that exist are atoms, molecules, and chemicals.* If, as atheists claim, religion is a human construct that has artificially created concepts such as "God," "revelation," and "commandment," then atheistic ideologies are human constructs that have artificially created concepts such as "values," "meaning," and "morality." That is to say, a system of values based on an atheistic world view is in essence another form of "religion" that has at its source nothing more significant than the imagination of its human creator.

Here is a pronouncement on the human moral condition by Christopher Hitchens, from his aforementioned lecture at Sewanee University:

> If it was true...that we are all part of a grand Divine design...What would it actually mean?...It would mean a regime of permanent supervision and surveillance over our lives and personalities... an inescapable authoritarian control...It would be like living in a celestial North Korea...why would you want such a thing to be true?! What a hideous realm of permanent total inescapable unfreedom [*sic*] is being proposed to you...the struggle to throw off this servility is the precondition for liberty, whether personal, intellectual, or moral.[8]

Why is Hitchens so disturbed about a world with a grand Divine design? Is it because it implies that there is a real, transcendent purpose to our existence? Why does he find it so unnerving to think of a world with Divine supervision and authority? Is it because it would mean we would actually be held accountable for what we did with our lives? *Real Purpose? Accountability for our actions?* Gee, there goes all the fun. It almost sounds like what happens when we go to work. I guess with a grand Divine design, we'd be stuck having to conduct ourselves...like responsible adults.

In the first chapter, I pointed out the profound disconnect in Hitchens'

---

* In other words: We are living in a material world, and he (Baggini) is a material boy.

thinking process; this seems to be another example. Did anyone notice which government Hitchens describes as the epitome of authoritarian repression? North Korea, *an officially atheistic regime.* Mazel Tov! Now I understand the "moral necessity of atheism." Perhaps the reason why the government of North Korea is so murderously repressive is because they haven't seen Hitchens' lecture. If they did, they would finally find out that "the struggle to throw off this servility [of a grand Divine design] is the precondition for liberty, whether personal, intellectual, or moral."

Hitchens' eloquent emphasis on the nobility of personal liberty effectively obscures his evasion of the obvious — that throwing off the yoke of Divine authority simply means that one is "free" to do whatever one pleases — as long as the individual can psychologically reconcile himself with the behavior, and as long as he can get away with it. This is exactly the world that William Golding portrayed in the novel *The Lord of the Flies*: a group of human creatures with complete "freedom" who ultimately are responsible to no one and accountable to no one but themselves. The Talmudic phrase in Aramaic that describes this very old ideology is "*Leis din, v'leis dayan* — There is no judgment and there is no judge."

A moral thought problem for Hitchens — Imagine a large cruise ship sailing through the Pacific Ocean. On board in one of most luxurious suites is a ten-year-old boy along with his guardian. The young boy is the sole heir to a ten-billion dollar fortune. Around his neck he wears a special necklace with the code that gives him access to the Swiss bank account with all the money. In one of the more inexpensive rooms is a middle-aged man, who from his meager salary as a janitor has saved up money for years to take the cruise. The ship sinks in a storm, and the boy and the janitor are the only survivors, washed up on a small uncharted island. The janitor notices that a rescue ship has just appeared over the horizon and is slowly approaching. He makes the following calculation: **(a)** No one else in the entire world knows that this boy survived; **(b)** if we are both rescued, the boy returns to his ten-billion dollars and I go back to being a janitor; **(c)** if I kill the boy, take his necklace, and bury him in the sand, no one will know what happened and I can have the ten billion dollars; everyone dies eventually, so what real difference does it make?

This janitor is not a sadist and certainly doesn't want the boy to suffer, so he creeps up quietly from behind, whacks the boy over the head with a heavy rock, and buries him in the sand.

What is our moral evaluation? For the believer the answer is simple and decisive: God is watching and judging and will hold him accountable. However, from the perspective of the atheist/materialist, inasmuch as not a sole being in the universe is aware of what happened, the following question must be asked: *Did a murder even occur?* For Hitchens, the answer is a clear *no. Leis din, v'leis dayan*. There is no judgment and there is no judge. There is only the shimmering, vapid mirage called "personal freedom."

The Incoherent Moral Philosophy of Michael Ruse

Michael Ruse is professor of philosophy at Florida State University. Some of his posts on the *Brainstorm* section of *The Chronicle of Higher Education* site on the subject of morality were the focal points of a sharp disagreement between him on one side and Dr. Jerry Coyne and Dr. Jason Rosenhouse on the other. In an article entitled "Scientism Continued," Ruse, an atheist, made the astounding claim that not only is pedophilia immoral, but that this moral principle is an *objective truth!*

> I want to say that what Jerry Sandusky [a convicted child molester] was reportedly doing to kids in the showers was morally wrong, and that this was not just an opinion or something based on subjective value judgments. The truth of its wrongness is as well taken as the truth of the heliocentric solar system.[9]

Where is this source of objective moral truth that Ruse seems to have discovered? What is the identity of this absolute moral authority with whom Ruse has been consulting? In an earlier article that appeared in *The Guardian* in March 2010, Ruse appears to contradict what he wrote above:

> Morality then is not something handed down to Moses on Mount Sinai. It is something forged in the struggle for existence and reproduction, something fashioned by natural selection...Morality is just a matter of emotions, like liking ice cream...now that

> you know morality is an illusion put in place by your genes to make you a social cooperator, what's to stop you from behaving like an ancient Roman [raping and pillaging]? Well, nothing in an objective sense.[10]

Here, Ruse is clearly stating that morality is purely *subjective*. It's not like he is the first thinker to come to this conclusion. It is glaringly obvious to theists. Ruse understands the dilemma quite well. A *subjective* system of morality is nothing more than a rickety shack with no foundation; it will collapse in the first good wind:

> But [morality] has to be, a funny kind of *emotion*. It has to *pretend* that it is not that at all! If we thought that morality was no more than liking or not liking spinach, then pretty quickly it would break down...very quickly there would be no morality and society would collapse and each and every one of us would suffer.[11]

This is essentially the same point made by Russell cited earlier. Is there an escape from this seemingly intractable problem? Ruse offers us his solution:

> So morality has to come across as something that is more than emotion. It has to appear to be objective, even though really it is subjective...Because that is what morality demands of us. It is bigger than the both of us. It is laid upon us and we must accept it, just like we must accept that 2+2=4.[12]

Let us offer a brief analysis of what Ruse has written so far:

- "Morality has to come across as something that is more than emotion. It has to appear to be objective, even though really it is subjective." If morality, in truth, is just an emotion, by what faculty do we make it into something more? In fact, in an atheistic/materialistic world, morality is nothing more and nothing less than the crackling of a bunch of electrical synaptic connections somewhere in the human brain. How could it possibly be anything more?! How can you make an emotion into something other than an emotion? How do you make something that is subjective "appear" to be objective? I have a suggestion: try "Presto" or "Abracadabra"!

- "It is bigger than the both of us....we must accept it, just like we must accept that 2+2=4." It's bigger than the both of us? Sounds suspiciously religious and God-like to me. It does however give me a wonderful idea how we can get people to "just accept it" as an absolute, mathematical truth: **In the beginning was Professor Michael Ruse; and Michael Ruse thus did speaketh: Let there be objective morality!**

In another *Brainstorm* article, entitled "The Nature of Morality: Replies to Critics," Ruse "clarifies" [?!] the issue further:

> If you place "subjective" in opposition to "objective," and mean by the latter something external, then clearly the kind of ethics I propose is subjective...but it is not subjective whether you think sodomizing little boys is right or wrong... My position is that evolutionary biology lays on us certain absolutes. These are adaptations brought on by natural selection. It is in this sense I claim that morality is not subjective.[13]

And back to the original article "Scientism Continued":

> So how do you justify moral claims? Some philosophers and theologians think you can do it by reference to so-called non-natural properties or perhaps the will of God. Others, and this includes me, think that perhaps morality has no objective justification in this sense...So what does this make of morality... As evolved human beings, the rules of morality are as binding on us as if we were the children of God and He had made up the rules.[14]

If all of this sounds incoherent it is because it *is* incoherent. I'm not the only one who thinks so. Some prominent atheist academics agree

The Moral Philosophy of Jerry Coyne and Jason Rosenhouse

On December 20, 2011, Dr. Jerry Coyne offered us a post on his *WhyEvolutionIsTrue* blog entitled, "Ruse goes after scientism again, but screws up on morality." Need I say more? We don't even need to cite anything from Coyne's article. Ruse himself wrote, "My most doughty critic, Jerry Coyne, says: 'While science can inform moral judgments, in the end statements about right or wrong are opinions, based on subjective value judgments.'" Much to my chagrin, I actually find myself enthusiastically agreeing with Jerry Coyne! Coyne also tells us that Dr. Jason Rosenhouse on *Evolution Blog* had already pointed out the glaring flaws in Ruse's approach very effectively. Rosenhouse echoes my own observations regarding the incoherence of Ruse's position:

> I can't follow this at all...He seems to be saying that Sandusky's actions are really and truly wrong because natural selection has programmed us to believe they are wrong. Can someone explain what I am missing? It sure looks like Ruse has contradicted himself here...Concepts of right and wrong differ among contemporary cultures. They also evolve over time. Can Ruse help us make sense of this? Can Ruse apply his methods to resolve any current area of moral controversy? Do appeals to psychology and natural selection help us resolve questions about abortion or homosexuality?... Ruse's essay was meant to establish that there are moral facts that we come to know by non-empirical means...To the extent that I understand what he is saying...*he has established neither that there are moral facts nor that he has some reliable, non-scientific means of determining what they are.*[15]

To which I can only add, Amen.

Let us sum up. Michael Ruse's proclamation that child-molestation is objectively immoral is based on an incoherent, self-contradictory atheistic moral philosophy. However, I do understand what is creating the cognitive dissonance in Ruse and frankly, I feel for him. It is the same terrible frustration expressed by Russell about wanton cruelty. Ruse is psychologically unable to accept that the only thing wrong with child-molestation is that *he doesn't like it*; or perhaps even more terrifying, that his abhorrence of child-molestation is nothing more than the results of his *societal conditioning*. He therefore has no choice but to take a *leap of faith* and declare his moral principles to be true with mathematical certainty and to be binding as if they were proclaimed by God himself.

To their credit, Coyne and Rosenhouse were not fooled by this gibberish, although this leaves us right back where we started from. By their own admission, in an atheistic world there is nothing inherently wrong with pedophilia or anything else for that matter. No one has stated it more clearly than the previously cited professor emeritus of philosophy at the University of New Haven, Dr. Joel Marks:

> The long and the short of it is that I became convinced that atheism implies amorality; and since I am an atheist, I must therefore embrace amorality. I call the premise of this argument "hard atheism"...a "soft atheist" would hold that one could be an atheist and still believe in morality. And indeed, the whole crop of "New Atheists" are softies of this kind. So was I, until I experienced my shocking epiphany that the religious fundamentalists are correct: without God, there is no morality. But they are incorrect, I still believe, about there being a God. Hence, I believe, there is no morality.[16]

If we are not accountable to a higher power for our actions, how we choose to behave becomes a question of "am I psychologically able to jettison the societal conditioning to which I have been subjected?" Please ask yourselves the following question: If one concludes, like Dr. Marks, that there is no such thing as right or wrong, as moral and immoral, *and* one has the desires of a pedophile, how would one then view the actions of Jerry Sandusky? If there is one thing we have learned from the bloody history of mankind, and particularly from the events of the twentieth century, it is that there is *nothing* that human beings are not capable of doing.

Ruse and I do agree on one thing. Once the atheist realizes that all of his noble moral principles are nothing more than subjective feelings — "no more than liking or not liking spinach" — then "pretty quickly it would break down...very quickly there would be no morality." Actually, it's not that there would be no morality. It's just that the moral values would change according to societal whim.

Dr. Sam Harris, Atheist author and Neuroscientist

Sam Harris is the author of a best-selling atheist manifesto entitled *The End of Faith: Terror, Religion and the Future of Reason.* In 2005 (the year after its publication), his book won the PEN American Center/Martha Albrand award. In 2006, Harris authored *Letter to a Christian Nation*, which one reviewer described as a "rejoinder to the criticism his first book attracted." Harris takes a rather novel approach in his attempt to overcome the insurmountable dilemma faced by all atheist thinkers: the problem of "the subjectivity of ethical values." He claims that *science* will provide the answers:

> *There will probably come a time when* we achieve a detailed understanding of human happiness and of ethical judgments themselves at the level of the brain...*There is every reason to believe that* sustained inquiry in the moral sphere will force convergence of our various belief systems in the way that it has every other science...[17]

In the passage above, Harris provides us with two powerful reasons to believe that science holds the key to solving this dilemma: (a) "*There will probably come a time when...*" and (b) "*There is every reason to believe that...*" How is that for compelling, rational evidence? I didn't realize how easy it was to establish a fact. You don't need to present evidence, logic, or construct a reasonable argument. All you have to do is preface your agenda with "there will probably come a time when (fill in the blank)" or "there is every reason to believe that (fill in the blank)" and you're good to go! Let's try it: *There is every reason to believe that there will probably come a time when* even Sam Harris realizes that intractable philosophical problems cannot be solved by hand-waving and flimsy pronouncements!

Not only does Harris present us with the totally unsupported article of faith that science will settle all metaphysical questions regarding morality, but in *The End of Faith* he also indicates that a God-based view of the world necessarily includes a distorted perception of morality and ethics. Harris quotes the aforementioned philosopher Bertrand Russell: "There is something odd about the ethical evaluations of those who think that an omnipotent deity...would consider himself adequately rewarded by the final emergence of Hitler, Stalin, and the H-Bomb." Harris then offers his commentary on this statement by Russell: **"This is a devastating observation, and there is no retort to it.**"[18]

I have no objection to Harris raising the challenge that evil and suffering pose to the believer. Theologians and religious philosophers have pondered the question for millennia. It is so important that an entire book of Scripture — the Book of Job — is devoted exclusively to this subject. What I find objectionable is Harris' comment that "this is a devastating observation, and there is no retort to it." In Harris' mind, pointing out the crimes of Hitler and Stalin and the existence of H-Bombs conjures up images of vanquished theologians and clergy helplessly thrashing in the mud, while cultivated, broad-minded freethinkers like him continue climbing the path to enlightenment. It's as if Harris actually believes he was the first person to think of the question.

Is it really true that Russell's observations about Hitler, Stalin, and the H-Bomb are so "devastating" that the believer has "no retort?"

Perhaps Harris and his fellow skeptics should consider the following:

- **Hitler** — "Struggle is the father of all things...he who wants to live must fight, and who does not want to fight in this world where external struggle is the law has no right to exist"[19](*Mein Kampf*). "The stronger must dominate and not blend with the weaker...only the born weakling can view this as cruel...for if this law did not prevail, any conceivable higher evolution of organic living beings would be unthinkable" [20](ibid.).
- **Stalin** — An atheist who was one of the most evil men who ever walked the face of the earth.
- **H-Bomb** — In the world of the atheist, who are the aristocrats, the priestly class, the Brahman caste? The rational, logical scientists. Who created the H-Bomb? *Scientists*!

Let us sum up: **Hitler** — Darwinist par excellence; **Stalin** — Atheist mass murderer; **H-Bomb** — Created by the elite of the atheistic world, i.e., scientists. *That,* Dr. Harris, is a devastating observation and there is no retort to it.

Amorality and Survival of the Fittest

The amorality that is inherent in a non-theistic concept of values can be seen clearly in the following statements by a number of well-known atheists. The only inescapable and inviolable rule of existence is "survival of the fittest."

H.G. Wells (1866–1946), British author of *The Time Machine*, *The Invisible Man*, and *War of the Worlds*:

"The men of the New Republic will not be squeamish, either, in facing or inflicting death...they will have an ideal that will make killing worthwhile...they will contrive a land legislation that will keep the black, or yellow, or mean white squatter on the move...this thing, this euthanasia of the weak...is possible...it will be permissible, and I have little or no doubt that in the future it will be planned and achieved."[21]

Wells passionately advocated the implementation of the evolutionary principle of survival of the fittest by killing off "unfit" humans. Wells, who some called the "father of Science Fiction," also seemed to have a prophetic streak in his soul. His noble vision of euthanasia was planned and achieved by the Germans less than forty years after he wrote this in 1902. In Nazi Germany, the Darwinian inspired ideal was carried out with typical German meticulousness and efficiency. Scores of thousands of individuals identified as "useless eaters, persons devoid of value, worthless people, superfluous people, misfits, undesirables, cripples, schizophrenics, idiots, etc." were put to death in a plan carried out by a chain of mental hospitals, professors of psychiatry, and directors and staff members of mental hospitals.[22]

George Bernard Shaw (1856–1950), awarded the Nobel Prize for Literature in 1925:

"The moment we face it frankly, we are driven to the conclusion that the community has a right to put a price on the right to live in it... If people are fit to live, let them live under decent human conditions. If they are not fit to live, kill them in a decent human way."[23]

What a shame that Shaw wasn't a Nazi; he would have at least made sure that people in Auschwitz were murdered "in a decent human way."

Dr. Konrad Lorenz (1903–1989), awarded the Nobel Prize for Medicine in 1973:

"Just as in cancer the best treatment is to eradicate the parasitic growth as quickly as possible, the eugenic defense against the dysgenic effects of afflicted subpopulations is of necessity limited to *equally drastic measures*...when these inferior elements are not effectively eliminated... they destroy the host body as well as themselves."[24]

Gee, with Nobel Prize winners like these, who needs war criminals?

Havelock Ellis (1859–1939), physician and advocate for implementation of "survival of the fittest":

"Even when war is totally abolished, there is still a place in morality for killing...that is to say by killing the unfit...it is one of the unfortunate results of Christianity among us today...that we were led to reject infanticide... We know in the back of our minds that we only [reject infanticide] out of quaint superstition." [25]

It's really a quite natural flow and progression: We are animals; the animal kingdom is driven by natural selection and survival of the fittest; *therefore* let's use our evolutionary gift of intelligence to make our species *more fit* by eliminating the *unfit*. While certainly true that a non-believing individual is not *obligated* to agree with this approach, the logic is internally consistent.

Of course there are many non-theists who find this approach to be repulsive and embarrassing and whose nature is to be gentle, compassionate, empathic, and giving. That approach *appeals* to them, that's what they *prefer*. However, this has nothing to do with morality or moral principles. Keep in mind that dogs, cats, goats, cows, sheep, deer, etc., are all gentle and relatively harmless animals. Some "human animals" are like that too. Sheep are not harmless because of moral principles; that is their nature. To reiterate the point: There is nothing *immoral* about a cheetah killing a gazelle, and there is nothing *moral* about a beagle being friendly to children.

Josef Stalin, Mao Tse Tung, and Pol Pot (and other Communist despots), all atheists, caused more human misery and were responsible for the spilling of more innocent blood in a seventy-year period than all religious fanatics combined were able to cause, with their wars and atrocities, in the thousand years preceding Communism.

- **Mao Tse Tung (China)** — murdered fifty to seventy million people.
- **Stalin (Soviet Union)** — twenty to thirty million.
- **Pol-Pot (Cambodia)** — about two million (but in his defense we should point out that he only had the small country of Cambodia to work with).

In the amoral jungle of a Godless world, Stalin, Mao, and Pol Pot are the jackals, hyenas, and wolves; the animals of prey. On the other hand, Sam Harris, Richard Dawkins, and Daniel Dennet are the goats, cows, and field mice. Jean-Paul Sartre, French atheistic, existentialist philosopher, expressed this idea quite candidly: "It disturbs me no more to find men base, unjust, or selfish, than to see apes mischievous, wolves savage, or the vulture ravenous."[26]

A Pragmatic "Social Contract" Is Not Morality

For the non-theist, rules and regulations governing behavior are just a function of the "social contract" that different societies make in different ways at different times. Group cooperation is not morality. People in a village cooperate and make rules, or else life becomes intolerably stressful. That's being *practical*, not moral. However, there is no problem attacking, looting, and pillaging the *neighboring* village. In fact, to attack other villages and plunder their wealth can be *extremely practical*. Not only do you benefit from their wealth, possessions, women, and livestock, but it provides a sense of purpose and accomplishment. This is especially true if you outnumber them, have better weapons, and have stronger fighters.

The theist, on the other hand, believes that there is a *transcendent* source for his values, namely God. Ultimately, it is not really relevant if doing the right thing is "appealing" to him or not. The believer understands that he must subordinate his behavior, feelings, and even his innate natural tendencies to God's moral code. This is not to say that professing a belief in God alone will make a person moral and saintly; however, it does provide the only possible rational, philosophical, and spiritual basis for becoming moral and saintly. (The issue of atrocities and evil committed in the name of religion will be explored in the last chapter.) Even if the believer fails in his quest to live up to a divinely commanded morality because of human weakness, he still knows that God's moral commands are greater than, and more important than himself.

Moral values for the believer are not determined by personal preference, public opinion polls, and what the latest "progressive" trend happens to be. They are from the absolute being of the infinite, eternal God. There is no question that moral values can only be real and meaningful if their source is the transcendent God. In fact, only the believer can use the word "morality" in a meaningful and authentic way. The crucial issue for the believer that must be addressed (which I pointed out in Chapter 2 and which is well beyond the scope of this book), is how does one know that his values actually are what God has commanded?

Humanism: Another Red Herring

"Ethical" atheists will often preach what they label as "humanistic"* values. I would suggest that what prevents the humanist from committing murder with impunity is one of three factors:

A. He has no *desire or reason* to commit murder; for example, he might be very gentle and kind by nature;
B. If he does have a motive, he is afraid of getting caught; or
C. He has been conditioned by his society with God-based, Judeo-Christian values and is psychologically unable to jettison them.

Extremely Important to Note: There can also be a fourth reason: He actually does believe that there are transcendent values that supersede his own personal preferences. In other words, he's not really an atheist but does not yet realize it. He simply has not put two and two together.

It's easy to talk about humanism when you are feeling good. What does the humanist do when he is overwhelmed with burning anger, jealousy, lust, greed, etc., and he concludes that the most effective remedy to relieve the situation is murder? What would possibly stop him — besides

---

* **Humanism** is a philosophical and ethical stance that emphasizes the value and agency of human beings, individually and collectively... In modern times, humanist movements are typically aligned with secularism, and today "humanism" typically refers to a non-theistic life stance centered on human agency, and looking to science instead of religious dogma in order to understand the world. "Humanism," Wikipedia.com.

the psychological conditioning that I mentioned above — if he is certain he can get away with it? The hero of the atheist is the evolutionary biologist, like G. Gaylord Simpson. He has already told us that man is "purposeless...a sort of animal...a species in the order of primates." According to him, there is no essential difference between a man and a squirrel. If we kill squirrels when it suits us, then why not kill a man when it suits us? In the atheistic view we are in the previously cited words of Dr. Peter Walker, "...a carbon-based bag of mostly water, on a speck of iron silicate dust revolving around a boring dwarf star."[27]

What could possibly be wrong with killing a purposeless, directionless primate who anyway is just a "carbon-based bag of water" revolving pointlessly around a "boring dwarf star?" A moments thought will tell you that for the non-theist, nothing is intrinsically wrong with murder. If, for example, an Orthodox Jew, who truly believes that God has explicitly forbidden him to murder, is unable to control himself in the face of some enormous temptation and actually commits murder, it would be *despite* everything he believes in and stands for. If the atheist commits murder, it would be *entirely consistent* with what he believes in and stands for. By "entirely consistent," I do not mean that an inevitable consequence of his being an atheist is that he will commit murder. It is "entirely consistent" in the sense that there is nothing *intrinsically wrong* with committing murder.

One Man's "Deep Moral Feeling" Is Another Man's "Indigestion"

In order to escape the undeniable *amorality* of an atheistic worldview and the subsequent need to turn to a transcendent God for meaningful moral values, many ethical atheists promote the pathologically naïve idea that humans have some sort of *natural* morality inside their molecular structure somewhere. Hitchens, in his introduction to *The Portable Atheist*, puts it this way:

> The working assumption [of believers] is that we should have no moral compass if we were somehow not in thrall to an unalterable and unchallengeable celestial dictatorship. What a repulsive idea... The so-called Golden Rule is innate in us.[28]

In other words, we don't need moral direction or guidance from a Supreme Being; the voice of the "innate Golden Rule" will guide us and all moral issues will resolve themselves. It is hard to imagine a statement that more savagely violates human reason and common sense.

If Hitchens grew up in ancient Rome, wouldn't he be at the Coliseum along with all the other "good" Roman citizens, cheering wildly as gladiators killed each other and wild animals tore people to pieces? If he grew up in India, wouldn't he solemnly throw the widow on the funeral pyre just like any "good" Indian? If he was raised in the Hitler Youth, wouldn't he "know" beyond any doubt whatsoever that it is the destiny of the supreme, master Aryan race to conquer the world? If he were raised in Mississippi in 1930, wouldn't he be mugging for the camera at a lynching just like everyone else did in those good old days? Didn't the followers and advocates of these systems "innately" know the golden rule?

Doesn't everybody "innately" know that killing an unborn child is murder? Millions do know just that...and millions of others "innately" know that the noblest thing that can be done for women everywhere, and for all humanity, is to make abortion a Constitutional right. If the pro-life side is actually right, we are guilty of mass murder; if the pro-abortion side is right, all those opposed to abortion are cruel and heartless, stripping women of their most basic human rights. Both sides are certain that they "know" the truth. Both sides have the Golden Rule "innately!" Don't all decent, sensitive, compassionate, animal loving people everywhere "innately" know that bullfighting is a terribly inhumane, cruel sport? Can fifty million Spaniards and a hundred million Mexicans really be "innately" wrong?...***OLE!*** Clearly then, the voice of the "innate Golden Rule" gives radically different messages to different people in different places and times. Hitchens, not surprisingly, responds with the ultimate smokescreen of an answer to this simple and obvious problem:

> The so-called Golden Rule is innate in us, or is innate except in the sociopaths who do not care about others and the psychopaths who take pleasure from cruelty.[29]

It is critical to point out how Hitchens' desire to promote his agenda twists and distorts his reasoning process. In order to sidestep the *amorality* of atheism and the self-apparent truth that human moral sensitivities are notoriously subjective and fickle, Hitchens simply labels adherents of value systems that conflict with his own as "psychopaths" and "sociopaths." He would have us believe that no sane, psychologically healthy person would ever justify the murder of a helpless baby; after all, our "innate Golden Rule" would never allow it. Nothing could be further from the truth. In 1978, anthropologist Laila Williamson, of the American Museum of Natural History in New York City, reached the following conclusions after a study on the prevalence of infanticide:

> Infanticide has been practiced on every continent and by people on every level of cultural complexity, from hunters and gatherers to high civilization, including our own ancestors. Rather than being an exception, then, it has been the rule.[30]

I would add that infanticide is still a common practice in India and China, particularly when the baby is female. Are we to conclude then that all these people were, and are, "psychopaths" or "sociopaths"? Dr. Steven Pinker, in an article entitled "Why They Kill Their Newborns," directly addresses the issue of the psychological state of baby killers:

> Neonaticide, many think, could be only the product of pathology... But it's hard to maintain that neonaticide is an illness when we learn that it has been practiced and accepted in most cultures throughout history. And that neonaticidal women do not commonly show signs of psychopathology... Several moral philosophers have concluded that neonates are not persons, and thus neonaticide should not be classified as murder. Michael Tooley has gone so far as to say that neonaticide ought to be permitted during an interval after birth.[31]

If the "innate Golden Rule" is applied by different societies at different times in completely opposite ways, if it is unable to prevent the justification of the murder of millions of babies "on every continent" and "every level of cultural complexity," and is even unable to prevent a so-called "moral philosopher" at the University of Colorado by the name of Michael Tooley from concluding that murdering babies should be a legal option in the twenty-first century, **what exactly does this "innate Golden Rule" actually do anyhow?**

As long as we are on the subject, what about "psychopaths" and "sociopaths" like Plato, Aristotle, Seneca, Cicero, Wells, and Shaw, all of whom casually and matter-of-factly endorsed the practice of infanticide? Roman historian Cornelius Tacitus (whom we have no reason to believe was psychologically unhealthy) counted among the "sinister and revolting" practices of the Jews the fact that they considered it a "deadly sin to kill a born or unborn child." [32]

Did Hitchens not realize that a fundamental tendency of nearly all human beings (irrespective of their psychological health) is to experience a powerfully "innate" feeling that their own particular value system, *no matter what it may be*, is perfectly moral and good? Simply put, just about everybody is "innately" positive they are doing the right thing. ("All of a man's ways are proper in his own eyes..." Proverbs [21:2]) And when I say everybody, I mean *everybody*. The following is from a translation of a speech made by Reichsfuhrer-SS Heinrich Himmler (among the top three or four most powerful men in Nazi Germany), at Poznan, Poland on October 4, 1943, to a gathering of SS Officers:

> "I also want to mention a very difficult subject before you here, completely openly. It should be discussed amongst us, and yet, nevertheless, we will never speak about it in public... I am talking about the Jewish evacuation, the extermination of the Jewish people. It is one of those things that is easily said, the Jewish people is being exterminated. Every Party member will tell you: perfectly clear, it's part of our plans, we're eliminating the Jews, exterminating them, ha! [as if it's] a small matter.
> And then along they come, all the eighty million upright Germans, and each one has his decent Jew. They say: all the others are swine, but here is a first-class Jew. And none of them has seen it, has endured it. Most of you will know what it means when 100 bodies lie together, when there are 500, or when there are 1,000. And to

> have seen this through, and — with the exception of human weaknesses — to have remained decent, has made us hard and is a page of glory never mentioned and never to be mentioned.
>
> Because we know how difficult things would be, if today in every city during the bomb attacks, the burdens of war and the privations, we still had Jews as secret saboteurs...and instigators... We have taken away the riches that they had, and I have given a strict order, which Obergruppenfuhrer Pohl has carried out, we have delivered these riches, completely to the Reich, to the State. We have taken nothing from them for ourselves...he who takes even one Mark of this is a dead man. A number of SS men have offended against this order. There are not very many, and they will be dead men — without mercy! We have the moral right, we had the duty to our people to do it, to kill this people who wanted to kill us, but we do not have the right to enrich ourselves with even...one cigarette...with anything. That [right] we do not have. Because at the end of this, we don't want, because we exterminated the bacillus, to become sick and die from the same bacillus.
>
> I will never see it happen, that even one bit of putrefaction [*sic*] comes in contact with us or takes root in us...*But altogether we can say: We have carried out this most difficult task for the love of our people. And we have taken on no defect within us, in our soul, or in our character.*[33]

Kind of stirring, isn't it? Ethical atheists everywhere, including of course Christopher Hitchens, can take great comfort in the fact that Himmler did not ignore his Darwinian gift of a genetically encoded Golden Rule, even as he was slaughtering millions of people. He was very careful to "remain decent," to only do that which he had a "moral right" to do, not to forget the "love of his people," and in an extraordinary display of virtue and ethical sensitivity, he and his SS colleagues scrupulously avoided taking on "defects" in their soul or character. Rabbi Yaakov Weinberg once remarked, "Nobody ever let something as trivial as facts and logic interfere with his agenda; if the facts and logic don't fit, then the facts and logic will just have to fend for themselves."[34] When it came to his own moral philosophy Hitchens has left the facts and logic to "fend for themselves."

The "Moral Compass" and Himmler's Speech

In truth, when all is said and done, I agree with Hitchens; we *do* have an inborn "moral compass." It is regarding the significance, function, and source of this "moral compass" where our opinions and worldviews radically diverge. As a believer, my position is that moral values and our inborn "moral compass" have actual metaphysical existence. Although moral principles emanate from God and are spiritual, existing in time but not space, they are as much a part of our reality as gravity, chemical bonds, and electromagnetic forces. That is why I find it extremely significant that Himmler felt the need to morally justify his actions. Even Himmler could not totally escape the moral voice that God implanted in the soul of every human being.

What is clear, however, is that the inborn moral voice *by itself* is virtually useless — demonstrated by the wildly disparate value systems of different individuals and societies. A ship at sea on a stormy night can only navigate accurately towards its destination by homing in on the beacon from a lighthouse situated on the firm foundation of dry land. Unless our inner "moral compass" is guided to seek its moral objectives by homing in on an absolute and immutable system of God-given moral values and principles, there is literally no perversion or corruption imaginable that cannot be justified by our inner "moral compass." The overwhelming majority of those reading this book would certainly agree that Himmler perverted and abused his innate moral sense. However, the absolute, transcendent moral laws that he violated still remain eternally in place. From the perspective of the theist, despite the horrific deviations of Himmler from the moral path, the God-implanted moral sense in every human being and the God-given moral principles still retain their ultimate significance and function.

From the viewpoint of the atheist, however, Himmler's speech only underscores how hollow and insignificant this inborn "moral compass" is, no matter what its chemical, genetic, or evolutionary source. For the atheist, moral principles have no actual existence. They are products of human desire, human construct, and human imagination. Since all human preferences, including moral values, are purely subjective, this "moral compass" (whatever it actually is) will obviously lead an individual wherever he or she subjectively would like to go — including morally justified infanticide, or for that matter mass murder, committed for the noblest of reasons. When the atheist honestly confronts the fact that this "moral compass" can produce a Himmler and a Stalin on the one hand, a humanist atheist on the other, and everything in between, he must be struck with an obvious question: what difference does this "compass" really make? *All of them feel equally strongly that their behavior is morally justified and ethically noble.* I would take it a step further. In the atheistic worldview, it would be quite reasonable to conclude that the entire evolutionary purpose of our evolved inborn "moral compass" is to give us the ability to justify and feel good about any system of values we choose to follow, no matter what it may be!

Why Have the Greatest Mass Murderers Been Atheists?

Richard Dawkins, the undisputed leader of modern atheism, attempts to disassociate himself from the mass murder committed by atheist tyrants with the following: "The bottom line of the Stalin/Hitler debating point is very simple. Individual atheists may do evil things, but they don't do evil things in the name of atheism."[35]

He's absolutely correct. Nobody makes wars or commits mass murder in the name of atheism. The reason for this is the same reason that nobody makes a war in the name of hopscotch. In order to do "great" things, you need a "great" idea. You cannot inspire people to fight a war in the name of hopscotch because it is trivial. Atheism is the most trivializing idea of all; it implies that we are a bunch of purposeless, meaningless, glorified bacteria spinning around in space. In other words, the implicit message of atheism/Darwinism is that the human is to the lobster what the lobster is to the cockroach; and the lobster is to the cockroach what the cockroach is to the paramecium. How could you possibly arouse the masses to do *anything* in the name of atheism?

However, while the concept of atheism itself negates any form of greatness, being an atheist does not eliminate the *inner need of the human being to seek greatness.* Once an individual has thrown off divine moral restrictions, he is now free to use all his ambitions, talents, energy and passion to achieve "great" things, *without any moral boundaries whatsoever.* Wells, Lorenz, Shaw, and the Nazis did not advocate murdering unfit humans in the name of something as trivial as atheism. They advocated these policies in the name of the glorious cause of "creating a race of thoroughbreds," as Planned Parenthood founder Margaret Sanger so eloquently termed it, or building a "New Republic," or "perfecting the human species," or "creating a Master Race." It was their non-belief, however, that intellectually, psychologically, and spiritually allowed them to justify the use of any means necessary, without "the inescapable authoritarian control" of divine moral restrictions, to achieve their goals.

Stalin, Mao Tse Tung, and Pol Pot absolutely and literally applied the words of Lenin: "Our morality is entirely subordinated to the interest of the proletariat's class struggle,"[36] and those of Mikhail Bakunin, one of the founders of the Anarchist movement: "to [the revolutionary], whatever aids the triumph of the revolution is ethical; whatever hinders it is unethical and criminal."[37] While undeniably true that terrible crimes have been committed in the name of religious ideologies, abandoning an eternal, God-based moral system seems to create the one thing that is potentially even more dangerous: atheistic ideologies. Any number of non-theists seem to be in denial of this obvious truth. Some examples:

- **Bertrand Russell, *Selected Essays*, 1928** — "I hope that every kind of religious belief will die out...although I am prepared to admit that in certain times and places it has had some good effects. I regard it as belonging to the infancy of human reason, and to a stage of development which we are now outgrowing."[38] This remark was published in 1928. Atheist mass murderer Josef Stalin became Secretary General of the Communist Party in 1922.
- **Sam Harris, *The End of Faith*** — "Ideas which divide one group of human beings from another, only to unite them in slaughter, **generally** have their roots in religion."[39] Harris would do better in the future to avoid speaking in ***generalities,*** because ***specifically*** the greatest mass murderers have been his fellow atheists.

- **Carl Van Doren, professor of English literature at Columbia University, and Pulitzer Prize-winning author** — "The unbelievers, as I have read history, have done less harm to the world than the believers. They have not filled it with savage wars...with crusades, or persecutions, with complacency or ignorance. They have instead, done what they could to fill it with knowledge and beauty, with temperance and justice, with manners and laughter."[40] The "knowledge and beauty" of Stalin and the Gulag, the "temperance and justice" of Mao Tse Tung ("political power grows from the barrel of a gun"), and the "manners and laughter" of the great unbeliever Pol Pot as he soaked Cambodia with the blood of millions.
- **Emma Goldman, anarchist and social activist, *The Philosophy of Atheism*** — "To what heights the philosophy of atheism may yet attain, no one can prophesy. But this much can be predicted; only by its regenerating fire will human relations be purged from the horrors of the past."[41] I have no comment; I just don't know whether to laugh or cry.

In short, the world of the atheist is the amoral world of the jungle, the forest, the anthill, and the sea. If there is any doubt at all as to the inescapable amorality inherent in an atheistic worldview, consider Dr. Peter Singer, world-famous philosopher and intellectual father of the animal rights movement. Professor Singer, who is a native Australian, was named Humanist of the Year in 2004 by the Council of Australian Humanist Societies and is currently the DeCamp Professor of Bioethics at Princeton University. In an article entitled "Heavy Petting," Singer gave his stamp of approval to bestiality as long as the animal is not harmed in any way. In a videotaped interview with journalist William Crawley, Singer was asked if he thought that pedophilia was "just wrong." Singer became what might be described as mildly indignant and responded:

> I don't have intrinsic moral taboos. *My view is not that anything is just wrong*...You're trying to put words in my mouth...I don't think that this moral method of saying it's just wrong is a method we should rely on, neither in this case [pedophilia], nor any other.[42]

Singer went on to say that he is a "consequentialist [*sic*]." This is a clear case of obfuscation masquerading as ethical philosophy. Every mafia hit man is a "consequentialist." If there is a policeman around and he is afraid of getting caught (i.e., bad consequences) he will not commit murder. Seals at a circus show, waiting for their trainer to throw them a fish after they have successfully completed a trick (i.e., good consequences) are also "consequentialists." (Perhaps Singer should join the circus.) In light of the fact that Singer is a professor of *ethics* (!) at Princeton University, I feel compelled to record here my revised version of a stinging rebuke delivered by the late William F. Buckley Jr. (substituting "Princeton University" where Buckley originally wrote "Harvard University"): "I would rather entrust the government of the United States to the first 400 names in the Boston telephone directory than to the faculty of Princeton University."

In conclusion, let's review some of the pronouncements made by prominent non-theists about the relationship between moral values and atheism:

Peter Singer — "I don't have intrinsic moral taboos. My view is not that anything is just wrong."

Jean-Paul Sartre — "The existentialist finds it extremely embarrassing that God does not exist, for there disappears with him all possibility of finding values in an intelligible heaven."[43]

Dr. Joel Marks — "I have given up morality altogether! [I] have been laboring under an unexamined assumption, namely, that there is such a thing as right and wrong. I now believe there isn't...I experienced my shocking epiphany that the religious fundamentalists are correct; without God there is no morality...Hence I believe there is no morality...The long and short of it is that I became convinced that atheism implies amorality; and since I am an atheist, I must therefore embrace amorality..."[44]

Jeffrey Dahmer[*] — "I always believed the theory of evolution as truth that we all just came from the slime...if a person doesn't think there is a God to be accountable to then what's the point of trying to modify your behavior to keep it within acceptable ranges?"[45]

---

* Jeffrey Dahmer (1960–1994), a convicted serial killer who murdered seventeen boys and men in the years 1978–1991. He was murdered in prison.

Mike, "EvilTeuf," atheist blogger — "Atheism... doesn't carry any obligation to any kind of political or moral system. In that sense it is amoral...What the amorality of atheism entails is a lack of obligation to any system of morality...An atheist can have any system of morality he or she wishes...you should always remember that no morals are absolute and that you always have a choice."[46]

After reading the above, it becomes very difficult to understand what Christopher Hitchens could have been thinking when he called his lecture at Sewanee University "The Moral Necessity of Atheism." It seems that it should have been called: *No Morals Are Absolute, Nothing Is Intrinsically Wrong, and Religious Fundamentalists Are Correct — Without God There Is No Morality.*

This is not the end of the discussion, though. Many non-theistic philosophers have an intriguing response. They do not deny that atheism provides no basis for moral values; rather, they claim that *both theist and non-theist alike* are caught on the horns of the exact same moral dilemma. With this, then, we are ready to introduce what many non-theists consider to be one of the most potent philosophical weapons in their arsenal — an argument with the quaint Greek name of Euthyphro.

Endnotes

1 G. Gaylord Simpson, "The Meaning of Evolution," as cited by David Oderberg in *Real Essentialism* (Routledge, 2007), p. 241.

2 Brad Delong, *Brad DeLong's Grasping Reality Blog, http://delong.typepad.com/sdj/2012/11/sun-sets-but-sun-also-rises-jumped-up-monkeys-with-darwinian-heuristics-vs-angelic-reasoning-beings-with-direct-unmediated-a.html*

3 *The Notebooks of Samuel Butler* (Paperback Edition, Biblio-Bazaar, 2006), p. 48.

4 Cited in Prager and Telushkin in *The Nine Questions People Ask about Judaism* (New York: Simon & Schuster, Inc., 1986), p. 22.

5 Dr. Joel Marks, "An Amoral Manifesto, Part 1," *Philosophy Now*, Aug./Sep. 2010, http://www.philosophynow.org/issue80/80marks.htm

6 Baggini, *Atheism: A Very Short Introduction, p. 2.*

7 Ibid. p. 3.

8 Hitchens, "The Moral Necessity of Atheism" YouTube.com

9 Michael Ruse, "Scientism Continued," *The Chronicle of Higher Education, December 19, 2011, http://chronicle.com/blogs/brainstorm/scientism-continued/42332*

10 Ruse, "God is dead, long live morality," *The Guardian, March 15, 2010.*

11 Ibid.

12 Ibid.

13 Ruse, "The Nature of Morality: Replies to Critics," *The Chronicle of Higher Education, http://chronicle.com/blogs/brainstorm/the-nature-of-morality-replies-to-critics/42558*

14 See note 9.

15 Jason Rosenhouse, "The Basis for Morality," *December 19, 2011, http://scienceblogs.com/evolutionblog/2011/12/*

16 See note 5.

17 Harris, *The End of Faith, p. 175.*

18 Ibid., p. 173.

19 http://www.thedarwinpapers.com/oldsite/number12/Darwinpapers12HTML.htm (see footnote 13 at this website).

20 Adolf Hitler, *Mein Kampf*, Vol. I, Chapter 11.

21 H.G. Wells, *Anticipations of the Reactions of Mechanical and Scientific Progress upon Human Life and Thought*, 1902.

22 Frederic Wertham, *A Sign For Cain: An Exploration of Human Violence (Hale, 1968).*

23 George Bernard Shaw, *Prefaces*, (London: Constable and Co., 1934).

24 Dr. Konrad Lorenz, from *The Legacy of Malthus: The Social Costs of the New Scientific Racism*, A. Chase, (Alfred Knopf, 1980) p. 34.

25 Havelock Ellis, from an essay entitled "The Control of Population," cited in *Breeding Superman*, Dan Stone (Liverpool Univ. Press, 2002), p. 76.

26 http://citador.pt/quotes/citador.php?cit=1&op=7&author=19&firstrec=0; http://www.brainyquote.com/quotes/quotes/j/jeanpauls118277.html; http://thinkexist.com/quotation/it_disturbs_me_no_more_to_find_men_base-unjust-or/211342.html

27 Huberman, *The Quotable Atheist, p. 313.*

28 Hitchens, *The Portable Atheist*, Introduction.

29 Ibid.

30 Dr. Larry Milner, "A Brief History of Infanticide," *The Society for the Prevention of Infanticide* (1998), http://www.infanticide.org/history.htm

31 Dr. Steven Pinker, "Why They Kill Their Newborns," *The New York Times*, November 2, 1997, Magazine Desk Section, http://www.rightgrrl.com/carolyn/pinker.html

32 Cornelis Tacitus, *The Histories* 5:5 (the Jews), http://civilizationis.com/smartboard/shop/tacitusc/histries/chap18.htm

33 http://en.wikipedia.org/wiki/File:Himmler_Posen_Speech_-_Extermination_of_the_Jews_excerpt,_Oct_4,_1943.ogg (you can also hear the original recording of the speech)

34 Heard by the author in a lecture by Rabbi Yaakov Weinberg, of blessed memory.

35 Dawkins, *The God Delusion, p. 315.*

36 Leonard Roy Frank, *Webster's Quotationary*, Random House Information Group, 2001, p. 524.

37 Ibid., p. 247.

38 Ibid., p. 715.

39 Sam Harris, *The End of Faith: Religion, Terror, and the Future of Reason*, (New York: First Edition, W.W. Norton, Inc., 2004) p. 12.

40 Carl Van Doren, "Why I Am An Unbeliever," http://www.skeptically.org/thinkersonreligion/id10.html

41 Hitchens, *The Portable Atheist*, p. 133.

42 "William Crawley meets Peter Singer," http://www.youtube.com/watch?v=gAhAlbsAbLM&feature=related

43 Jean-Paul Sartre, "Existentialism is a Humanism," lecture (1946),http://www.marxists.org/reference/archive/sartre/works/exist/sartre.htm

44 See note 5.

45 Jeffrey Dahmer, from an interview with Stone Phillips on *Dateline, NBC,* November 29, 1994, http://scienceblogs.com/insolence/2007/10/evolution_made_me_do_it.php

46 "Where Do Atheists Get Their Morality From?" http://www.mwillett.org/atheism/moralsource.htm

Chapter 9

The Euthyphro Dilemma

As we discussed at length in the last chapter, in a universe without God there simply are no real, binding moral principles. Non-theistic philosophers reply that designating God as the moral lawgiver still does not solve the problem of discovering true moral principles. They contend that both the believer and atheist are stuck with the same philosophical/moral dilemma. Dr. Julian Baggini states this idea explicitly: "Of course, it can still be said that we can provide no logical proof that the atheist ought to behave morally, but neither can we provide such a proof for theists."[1]

The first part of Baggini's statement, that there is no logical reason for an atheist to behave morally, is obviously true as we have pointed out. However, what does Baggini mean when he writes, "but neither can we provide such a proof for the theists"? Why doesn't the theist, or believer in God, have a logical, rational reason to behave morally? If I believe that the infinite Creator of the universe commands me to behave in a certain way, isn't that a logical reason to behave morally? What Baggini is referring to is a philosophical challenge found in the works of Plato, in a dialogue entitled *Euthyphro*.

The Euthyphro Argument has been presented in various formulations. Bertrand Russell, in a work entitled *Why I Am Not a Christian*, presents one version. In a debate with Dennis Prager, professor Jonathan Glover used this argument to challenge Prager's assertion that morality must be God-based. It was also used by philosopher Kai Nielsen in a debate with Dr. William Lane Craig. It was even once written up in the religion column of the *Chicago Sun-Times*. The argument is well-known to theologians and philosophers of religion. At first glance, the argument seems very powerful. At least one blogger, Snaars.blogspot, July 2005, seems to indicate that it was instrumental in converting him from believer to non-theist: "When I saw this argument for the first time, I was stunned. I had been firmly of the belief that morality did depend on God's command! Eventually, I accepted the point of the argument [i.e., that morality, if it exists at all, has nothing to do with God's command]."

The Euthyphro Argument

I present below two versions of the Euthyphro Argument. The first is the one formulated by Julian Baggini, and is taken from his book *Atheism: A Very Short Introduction*:

> Plato made the point extremely clearly [sic] in a dialogue called Euthyphro... Plato's protagonist, Socrates, posed the question... **[A]** Do the gods choose what is good, because it is good, or **[B]** Is the good good, because the gods chose it? If the first option is true, that shows that good is independent of the gods (or God, in a monotheistic faith). Good is just good, and that is precisely why a good god will always choose it.
>
> But if the second option is true [i.e., it is good because the gods choose it], then that makes the very idea of what is good arbitrary. If it is God's choosing something alone that makes it good, then what is there to stop God from choosing torture, for instance, and thus making it good [or murder, or hatred, etc.]...to recognize this, however, is to recognize that we do not need God to determine right or wrong. Torture is not wrong just because God does not choose it...[2]

Here is a drier, more "mathematical" presentation of the same argument. Some readers might find this version easier to understand:

The Euthyphro Argument against the Divine Command Theory of Morality (DCT)

DCT states: Actions are wrong if, and only if, God commands us not to perform them.

1. EITHER: **(a)** God commands us not to steal, murder, lie, etc., because these actions are wrong; or **(b)** these actions are wrong because God commands us not to do them.
2. If **(a)** is true, then there is a standard of morality separate from God's commands and DCT is false.
3. If **(b)** is true, then either: **(c)** God has reasons for commanding us to avoid these actions; or **(d)** God has no reasons for commanding us to avoid these actions.
4. If **(c)** is true, then it is these reasons, whatever they are, that make these actions wrong, not God's commands, and DCT is false.
5. If **(d)** is true, then God's commands are arbitrary and effectively meaningless.
6. Thus, either DCT is false, or Divine moral rules are arbitrary and meaningless.

It seems that the believer is in the same patch of quicksand as the atheist. My contention though, as I will try to demonstrate, is that no matter how convincing it may sound at first, the Euthyphro Argument is really nothing more than a well-orchestrated piece of philosophical sleight-of-hand. It belongs in a Museum of Philosophy in the section that displays ancient philosophical fossils of ideas that used to walk the earth, but in reality crumbled into dust a long time ago. In truth, this argument was obsolete even before the time of Plato. It died with the appearance of Abraham and his introduction to the world of the revolutionary concept of Monotheism.*

The reason that the Euthyphro Argument is still talked about is a combination of two factors: **(1)** philosophers have not realized that the argument only applies to the pagan gods of Plato and the rest of the an-

---

* Actually, according to Jewish theology, Abraham *reintroduced Monotheism into the world.*

cient world, but not the One God of Monotheism; and **(2)** none of these philosophers ever bothered to formulate an actual meaningful *definition* of good, as opposed to simply presenting certain types of actions that they believe are good, and certain types of actions they believe are bad (for example: murder is bad, kindness is good). In the absence of these two factors the argument crumbles, as will be demonstrated.

The most effective and complete way to answer a philosophical problem is to show that the question never really was a question in the first place. It was all based on a fundamental misunderstanding of one sort or another. As we shall see, this is exactly the case with the Euthyphro Argument. Before I proceed, however, it is very important to restate and explicitly clarify the problem we are trying to solve. I have found that unless this is done many people become confused during the presentation.

A Clarification: What the Issue Is, and What the Issue Isn't

Please bear with me as I briefly review some of the things we have discussed. The believer argues that in a world without God, there are no objective moral truths. Every person, or group of people, will make up guidelines to live by as they see fit. The particular guidelines will reflect the tastes and opinions of different individuals and different societies at different times and places. For what human being or society has the authority to declare to all mankind what is right or wrong? Only a transcendent moral being can formulate and command moral truths that bind and obligate all humanity. Please note that even if this argument is true, it does not necessarily prove the existence of such a Being. It would only prove that without such a Being, there could be no objective moral truths. The atheist, of course, answers back using the Euthyphro Argument that *even if God exists* morality does not depend on God. Moral truths either exist independently of God or they are arbitrary and meaningless. To summarize: When we are discussing the Euthyphro Argument, the issue at stake is *not* whether God exists. The issue at stake *is* do we require the existence of God in order to have meaningful, absolute, objective moral truths? It is critical to keep this distinction in mind as we proceed to demonstrate the essentially flawed nature of the Euthyphro Argument.

A. Value Judgments Involve Comparisons

Invariably, value judgments involve comparisons. In other words, if I say someone or something is good or bad, it is in comparison to something that acts as a standard. For example, if I talk about a major league baseball pitcher as being "good," I am essentially comparing him to a theoretical model of what a pitcher is supposed to accomplish. The more strikeouts, fewer earned runs, more games won, etc., the "better" the pitcher. I can also make a comparison to great pitchers of the past or present.

Let's say I own a factory where I want to produce "excellent" MP3 players. At the end of the assembly line I will compare the MP3 players to a certain list of specifications: sound quality, frequency response, durability, etc. If the unit matches my specifications, I consider it an "excellent" system. Imagine the absurdity of a situation where I demand from my workers production of excellent electronics, but I give no specifications as to what excellence is. Without the standard of comparison, the whole concept becomes meaningless. When we talk about "good" or "bad" people, or "good" or "bad" behavior, there must be a standard to which we are comparing the people or behavior. If the standard is one that is subjective or arbitrarily chosen, the meaning of "good" or "bad" is purely subjective or arbitrary. It has no meaning in real or absolute terms. That is to say, it means whatever you want it to mean.

B. Contingent Reality and Actual Reality

Let us posit, however, the existence of the God of Abraham — the One transcendent, eternal, infinitely powerful, all-knowing Creator. (I remind the reader that we are not attempting in this chapter to prove either the truth of this God's existence or the truth of the implications of his existence which follow below. That is not the pertinent issue when discussing the Euthyphro Argument. We are positing here its truth for argument's sake in order to address the issues raised by the Euthyphro Argument.) Not only are we positing that this God brought the universe into existence, but that the source of the very being of the world and its continued existence is only God's continuous *will* for it to be and exist.

An approximate, acceptable analogy would be a person daydreaming and picturing a scene in his head of a beautiful, pristine, tropical beach. The sun is shining overhead. The waves are gently lapping along the shore. Palm trees are swaying in the gentle breeze. Suddenly the phone rings and this person must answer the phone and discuss an urgent business problem. Where did the beach go? What happened to the ocean, the sun, and the palm trees? A moment ago they were there, and now they're gone. Was the scene real? It had to be real in some sense; after all, he had vividly experienced the scene in his head. The answer is that the beach was real. It just did not exist on the same *plane of reality* as the person himself and the world around him. It was a *contingent* reality — a reality that was totally and completely dependent on the will of the person thinking or dreaming about it. The beach had no independent reality of its own. When this person "withdrew" his will, or in this case simply became distracted and turned his attention elsewhere, the beach ceased to exist.

Although certainly not the exact same thing (the analogy is a tool to relate in some way to the concept) we are positing that the universe came into existence and continues to exist because this God sustains it with his will. Our existence is contingent; we do not have *independent* existence and reality. This is the meaning of the phrase in the Jewish morning prayer service: "He renews every day *constantly* the act of creation." God created the world by his will and constantly wills it to continue to exist.

This God is not like us. This God in *actuality and reality* exists. Nothing caused or created him. *We* are created, along with time, space, matter, and energy. God, of course, is none of these. His existence is not at all like ours. God is Creator, we are creature. God ***is***; we are ***caused***. Nearly all human beings at one time or another are overwhelmed by a feeling of the "unreality" of our existence — that it seems like a dream; that our whole existence is just a "thought inside our head." I would suggest that these lurking feelings are our awareness at some primal level *that there is a plane of reality much greater than our own.* (Again, I am not trying to prove the truth of this idea here; I am just trying to explain what it is and its implications.)

C. "I Can't Get No Satisfaction" — What It Means that I Want to "Be" Somebody

Rabbi Yaakov Weinberg has stated in this vein that "the greatest need of a human being is to *be*."[3] What does it mean when people say, "I want to *be* somebody," or "I want *more* from my life?" Give a cow hay, some shade, and another cow for company, and it will be happy forever. Human beings are always looking for *more*. How many young and even not-so-young people "go on the road" to find this elusive something? What is there inside of us that elicits the almost universal identification with the adolescent Holden Caulfield in *The Catcher in the Rye* as he experiences the profoundly aching, bittersweet wonderment of self-discovery? Why do we ask ourselves who we are? What am I? Why am I here? It is not enough just to eat and breathe. That may satisfy those in the animal kingdom, but not us.

As human beings, we constantly need and seek something besides the simple fact that that we are alive to justify our existence. This justification can be found in religious or ideological commitment, a particular career, a challenging vocation, raising a family, running a marathon, receiving awards and honors, achievement, success, etc. Some discover and experience this by a deep contemplation and connection to the beauties and wonders of the natural world. I would suggest to you that this "looking for more," the burning desire to "be somebody," the excitement we feel when our existence is confirmed by seeing ourselves in the newspaper or television, is really about the search for *actual being and existence, and we intuitively know that on our own we don't have it*.

We seek a connection with something beyond ourselves and greater than ourselves to fulfill this need. From the perspective of Jewish/Monotheistic theology, the only way a human being can find *true* being and existence is to "attach" himself to this God and his absolute, eternal, infinite, and *actual* being and existence. If the created, contingently existent human can emulate, come close, and attach himself to the actual and real existence of God, his Creator, the human being in actuality and reality becomes *more*. He can "share," as it were, in God's actual being. *If this God exists*, then the closer I am to God and the deeper relationship I have with

God, the more I am in absolute terms. I repeat: I am, in objective reality, *more*. I am closer to the absolute, infinite, eternal existence that *is* God's being. Not only is there not anything more or greater than that, there isn't anything else period. He is the only true existence.

Perhaps the most profound attachment and intimacy one can have with God is to "be like him." Why do we want to be independent and feel relentlessly goaded by a deep compulsion to make it on our own? Why do we feel *less* if we are dependent? Every parent has experienced even a toddler's fierce drive to be independent. What is the tremendous pleasure of individual achievement and standing on our own two feet? To be independent *is to be like God himself*. God is absolutely independent. His existence and being depend on nothing at all outside himself. He needs nothing from anyone else. I would suggest that we have been created with an innate drive to form a relationship with God, by striving to be like God. That is why a young child passionately and sometimes almost violently proclaims, *without anyone having taught him*, "I want to do it myself!"

- Why is it that we agree that to be "strong" is better than to be "weak"? (I am not necessarily referring to physical strength, although generally we admire that also.) God's strength is infinite. God's will is unfaltering and unstoppable. The stronger we are, the more like God we are. The more willpower we exhibit, the more like God we are.
- God's power is without limits. The more powerful we are, the more like God we are.
- God's knowledge, awareness, and understanding are infinite. The more knowledge and understanding we have, the more like God we are.
- Since God needs nothing, the whole creation is a "gift" from God to his creatures. God, in this sense, is the ultimate "giver"; he only gives and gets nothing in return. What could anyone possibly give him? "Giving" is being like God. The more we give, the more we are like God.
- In some sense the greatest emulation of God is to be creative; God is *the* Creator. The pleasure and exhilaration we experience when we feel strong, independent, powerful, knowledgeable, giving, and creative is the pleasure of being Godly. It is the ecstasy of "true being."

This then is the only possible meaningful definition of "good" and "bad." The closer I come to the ultimate, infinite, actual being of God himself, the better I am. The further I move away from that absolute existence and reality that *is* His being, the worse I am. Connecting with this God means connecting with what ultimately is the only thing that is real; distancing oneself from this God means connecting to fantasy, hallucination, non-reality, i.e., nothingness. "Good or bad," "moral or immoral," are simply different ways to refer to closeness to, or distance from, God. Outside of this context, moral and immoral can only have purely subjective and/or arbitrary meanings.

It therefore follows that when this God commands us to "love your neighbor," it is not because it is "good" in the way non-theists (and many others) use the word, which is simply some vague notion of performing pleasant deeds. With his commandments, God offers us a *relationship.* He is saying to us: If you want to emulate Me, have a deeper relationship with Me, and attach yourselves to my absolute, infinite Being, then "love your neighbor." If you don't, you will be further away from Me and my absolute, infinite, true Being. "Thou shalt not murder" means that if you commit murder, you will be very, very far away from Me. The more you approach Me, the more of a "good" and "moral" person you are. If you choose to act in a way that moves you away from Me, you are "bad," "immoral," and "evil." The significance of moral values can only be understood in the context of, and as a function of, a deep relationship with God. Moral principles and commandments are analogous to a map that leads us to this relationship. The point is not *morality* or *commandments* per se. The point is to come close to God. The point is the relationship itself.

If this is still unclear, perhaps the following analogy will help. It is a very "good" thing for a husband to buy his wife a special gift on their anniversary that shows how much he cares for her and loves her. The reason why it is considered "good" is not because there is something inherently meaningful in shopping for gifts (I am aware that in certain circles that is a controversial statement). The "goodness" is because it strengthens the relationship between husband and wife. The "goodness" or "badness" in this context is whether or not the action brings them closer together or creates a distance between them. The pleasure of the relationship needs no justification or explanation; it justifies itself. In a similar way, following or disobeying God's commandments is only "good" or "bad" in the sense of the effect these actions have on one's relationship with the Creator. The pleasure and ecstasy of true being, of a relationship with our Creator, needs no justification or explanation — it justifies itself.

The source of God's authority in this matter also becomes apparent. *Only God Himself* knows what a human being must do in order to come close to God and what actions will cause a human being to become distant from God. Free will in this context, then, will ultimately boil down to this: will one choose to be "more" and come close to the infinite, eternal God, or will one be "less" by choosing non-reality and distancing oneself from God? (In a general way, distancing oneself from God usually takes the form of choosing to live as part of the animal kingdom.)

The Euthyphro Argument Vanishes into Thin Air

We can now see that in the context of Jewish theology, the Euthyphro Argument breaks down completely:

- Is "loving your neighbor" good simply *because* God commands it? Obviously not, that would make it arbitrary.
- Does God command "love your neighbor" because it connects to, or reflects some independent standard or concept of "good"?

No, it is neither of these. God commands us to love our neighbor so that we can choose to have a *relationship* with him, so that we can attach ourselves to his infinite and actual being; the relationship with God is *the* good. *If this infinite being we call God actually exists*, we have a real standard to determine a meaningful concept of moral truths. The standard is closeness to God, and the actual and absolute existence that is His being. This is what the Psalmist means when he proclaims, "To me, closeness to God *is* good," (Psalms 73:28). On the other hand, if this God does not exist, we are left with nothing but 100% subjective human tastes, preferences, opinions, and social mores. Subjective human value systems, even if they come from a so-called "Professor of Ethics" at Princeton University, have no meaning at all outside of the heads of those who follow them.

It becomes clear then that the Euthyphro Argument as a challenge to Judaism/Monotheism is nothing more than philosophical smoke and mirrors. The only reason it has some superficial appeal at all is because the word "gods" is used, giving the impression of some authority above human beings. Plato's original argument, of course, involved the pagan gods of Greece. However, a moment's thought will tell us that pagan gods have no more moral authority or moral credibility than mortal humans. A pagan god is simply a human being projected to a large scale. He's just bigger, stronger, lives longer, and can even throw a few lightning bolts when needed. Pagan gods are no different than The Incredible Hulk, The Flash, or Superman (who, as the old TV show told us, had "powers far beyond those of mortal men!"). Formulating the Euthyphro Argument using pagan gods is exactly the same thing as saying: *Does Superman command it because it is good, or is it good because Superman commands it*? The moral proclamations of Superman have no more or less significance than the moral proclamations of Zeus, Mick Jagger, Jerry Seinfeld, Jay Leno, or for that matter, any of the approximately six or seven billion other individuals living on this planet.

When stated this way, it becomes obvious how misguided and mistaken the whole argument was to begin with. What did you expect? Of course, pagan gods, superheroes, rock superstars, Jewish comedians, and even wildly successful talk-show hosts (just like everyone else) can only tell us their totally subjective views on values, or manufacture them arbitrarily.

Not so with the God of Abraham, the One God. The God of Monotheism is not a human being projected on a large scale. He is above time and space. He is above the physical. He is even above the spiritual. He *created* the spiritual. He is, as Rabbi Yaakov Weinberg has put it, "so totally and completely *other* than we are."[4] With the existence of the One God, greatness, goodness, meaning, morality, and eternal existence lie in front of us. They are within our grasp if we choose them. Without God, in the utterly empty void of the atheistic world, we are left with nothing but bleak despair (unless of course one creates a comforting fiction to live by), as expressed by the American novelist T. C. Boyle: "I am an atheist and a nihilist...I believe in nothing. And it causes me tremendous despair and heartbreak...there is nothing between us and the naked howling face of the universe. Nothing."[5]

I see no escape from the conclusion that atheism stands for nothing, signifies nothing, and affirms nothing except for one thing: all the moral aspirations and yearnings of the advanced primate we call a human being are nothing more than a cosmic joke...and not a very funny one at that. The choices before us are clear: we will either seek a transcendent moral law to which we will all submit, or we will seek our own personal and societal indulgence. If we turn to the One God in our quest to create a moral and just world, we have a fighting chance; if not, in my opinion, we are doomed to spiral into the amoral hell of the man-made human jungle.

End Notes

1 Baggini, *Atheism*, p. 56.
2 Ibid., p. 38.
3 Heard by the author in a lecture by Rabbi Yaakov Weinberg, of blessed memory.
4 Ibid.
5 *The Quotable Atheist*, p. 50.

Chapter 10

The Source of the Inborn Moral Imperative: The One God

Does Morality Have Actual Existence?

When all is said and done regarding God, morality, and atheism, we are left with only two intellectually consistent options:

a. God does not exist and there are no moral boundaries whatsoever; only subjectively agreed upon, non-binding, pragmatically driven "social contracts."
b. Morality is real and actual, and its source (as we discussed in Chapter 9) is the absolute, infinite, and eternal being of God himself.

If there were a way to confirm that *morality* was not just an artificial, imaginary human construct, but has actual metaphysical (beyond the physical) existence, we would know beyond a reasonable doubt that God exists. There would be no other possible source for such a phenomenon, as will be explained below.

Transcending the Need for Life Itself

We will start by carefully examining the human need for self-esteem. This examination will lead us to a clarity regarding the metaphysical reality of our moral drive and to its source.

We all have a deep inner need to see ourselves as important, valuable, and significant. According to Dr. Abraham Twerski, Director of the Gateway Rehabilitation Center (Aliquippa, PA) and a world-renowned expert on alcoholism and addiction, issues of self-esteem are at the root of almost every psychological problem that are not the result of a purely biological cause or chemical imbalance.[1] Whether we express it as "feeling OK about ourselves," "having a positive self-image," or "a sense of self-worth," the human need for self-esteem is, along with our need for meaning, the most powerful of all human drives. Under certain circumstances, our need for self-esteem transcends our need for life itself. Below are a few examples that illustrate this principle.

- Your family home is attacked by terrorists and you are faced with the following two choices: (a) run quickly out the back door and escape, leaving your wife and children at the mercy of the terrorists, or (b) risk your own life by shooting at the terrorists in the front of the house, allowing your family to escape out the back. Which would you choose? There certainly are people who don't care about anyone and would save themselves, but for the majority who would fight (I would think an overwhelming majority), for what reason are they prepared to die defending their wives and children? For the simple reason that most men simply could not (and would not want to) live with themselves if they fled and abandoned their loved ones. They would see themselves as completely worthless cowards, i.e., they would experience a total loss of self-esteem. They would literally risk their lives and die, if necessary, rather than live with the self-loathing that would result from the knowledge that they had not done everything within their power to save their families. (This judgment and assessment might not be made at a conscious level, but the split-second decision to fight would ultimately be based on these factors.) Although for the sake of simplicity I used the example of a man defending his family, it goes without saying that this principle would apply to a woman in the same situation.

- An elected official is caught on camera taking a bribe, selling classified information to a foreign government, or buying child pornography. The video is shown on all the major networks and uploaded to YouTube. This humiliation is simply a public stripping of the person's self-worth. If this official committed suicide, all of us reading the news would understand why he did it. We are prepared to die under certain circumstances rather than face a loss of self-esteem. The agonizing feeling of worthlessness makes life unbearable.
- In a neighborhood with street gangs, a gang member is insulted publicly in front of a group of friends, or even worse, in front of a group of girls. It is not at all unlikely that this gang member pulls out a gun or knife and attempts to maim or kill the person who insulted him. In his mind the very essence of his identity and sense of self-worth is being attacked and he reacts as if his very life had been threatened. This very same issue is usually the spark that ignites barroom brawls. Up until relatively recent times, a personal insult was grounds for a duel to the death among the aristocratic levels of society.
- In the 1982 award-winning film *An Officer and a Gentleman*, the character played by David Keith is betrayed by his fiancé. He checks into a motel, swallows the engagement ring, and hangs himself in the motel bathroom where he is discovered by his friend Zack (Richard Gere). The cruel and heartless manner in which he is rejected sets the stage for this dramatic moment in the film. The only reason this scene works so effectively is because we understand it is possible to literally kill someone by destroying their sense of self-worth. Along with the loss of self-esteem comes the loss of the desire to go on living.
- Imagine you are on a small cruise ship in the Mediterranean. Suddenly, terrorists board the ship and tie up all fifty passengers on deck. For some odd reason the leader of the terrorists takes a liking to you and offers you a deal to save your life. He hands you a hunting knife and says that if you kill all fifty passengers with this knife, he will set you free. There certainly are people who would not only take him up

> on this offer but would do it with glee. Thank God, I have been fortunate enough in my lifetime not to have met them. However, it is safe to say that nearly every single person reading this book would not be able to murder fifty people with a knife, even in order to save his own life. Why? What could possibly be more important than one's own life? Remember, when you refuse, you will die and never experience all the wonderful things life has to offer. No more beautiful sunsets, no more glorious spring days. No more happiness, no more family, no more love. What internal force is compelling you to give it all up? The answer is obvious. Living with the knowledge that you murdered fifty people would not be a life. You would rather be dead than have to live with yourself as an evil person. Seeing oneself as "evil" is another way of describing the complete disintegration of self-worth.

Although the need for self-esteem can be explored at many different levels, for our purposes the following statement is quite accurate. A person acquires self-esteem by *being* and *doing* those things that he considers to be good, meaningful, valuable, significant, or moral. Leaving aside issues of psychological pathology, it does not matter if those things that he considers to be good and valuable are objectively true or not. As long as the individual considers them to be true, generally speaking, they will nurture his sense of self-worth. It is self-evident that atheism and Christianity cannot possibly both be true. However, as long as atheists and Christians are firm in their convictions, loyalty to their respective ideologies will feed their sense of self-esteem.

It makes no difference what you consider to be good, moral, meaningful, or significant. Whatever it is that *you believe* fits into these categories — being popular with members of the opposite sex, getting high grades, success as an athlete, being a great warrior, acquiring wealth, being the toughest and most feared member of a gang, living an idealistic or spiritual life, being faithful to your religion — will be critical factors in determining your sense of self-worth and self-esteem. If I act in a way that *I myself* consider "immoral" or against my own values, I will suffer corresponding damage to my self-esteem. Pangs of conscience, guilt, feeling like a low-life, a nobody, a loser, or some other term all describe this shared human experience.

The amazing and fascinating conclusion that emerges from all this is that while there may be passionate and even violent disagreements on what actually is "good," *everyone* agrees that we need to "*be good*." Whether it is the German or Japanese soldier fighting for the Emperor or Fatherland, or the Allied soldier fighting for freedom from tyranny; whether it be the evangelical pastor denouncing sinners, or the atheist denouncing the pastor; whether it is the idealistic doctor in a third-world country, or the gang member standing up for his machismo; whether it is the man defending his family, the betrayed lover who commits suicide, the humiliated politician, the kid who spends hours on the basketball court dreaming of making it in the NBA, or everyday people trying to do what they believe is right; every single one of them is driven by the same innate need for self-worth and self-esteem. The inescapable need is there; it's just a matter of which values or value system you will use to "fill in the blank."

We Are Born with the Drive for Morality

It is clear we are born with both the innate need for self-esteem and a moral drive, which are two sides of the same coin. Allow me to explain. The youngest babies are capable of processing certain all-important signals they receive from their parents or caregivers. Loving, caring behavior is processed by the infant as "I am loved, cared for, and important," i.e., *I am OK*. If the parents convey the opposite, the infant processes that information as well, that *I am not OK*.

Dr. Lillian Katz, professor emeritus of Early Childhood Education at the University of Ill. Champaign-Urbana, and director of Education Resources Information Center, wrote: "The foundations of self-esteem are laid early in life...when adults readily respond to [infant's] cries and smiles, babies learn to feel loved and valued."[2]

The infant certainly does not consciously understand what is happening, but there are built-in "receptors" that respond to, process, and store "you are OK" messages and "you are not OK" messages. In the younger stages of growth, it is the parents who are the primary formers of the child's self-image. As the child grows older, peer relationships also become an important factor: "As they grow, children become increasingly sensitive to the evaluations of their peers."[3] General societal influences also kick in as the child develops and understands more about the world they are living in, "when children develop stronger ties with their peers… or around the neighborhood, they may begin to evaluate themselves differently from the way they were taught at home."[4]

At some point in the development of a human being, an almost miraculous transformation can occur. In what sometimes might be described as an epiphany, this person realizes:

> *I don't want to see myself as "good" or "OK" just because my parents, friends, teachers, and society approve of me and tell me I'm OK. I refuse to continue on my path like a nameless sheep in the herd; I refuse to be just "another brick in the wall." I want to be truly "good." I want to feel a sense of self-worth because I'm doing the things that I feel are truly meaningful and valuable.*

With this realization, the individual has arrived at a psychological and spiritual crossroads. Different people can make radically different choices at this crucial instant.

The Same Drive Causes One Person to Embrace Religion and Another to Reject It

At this point one person chooses to become deeply religious. Another heads in the opposite direction, abandons the faith of his youth, and embraces the "truth" of skepticism and atheism. At this point one person decides to leave his or her little home town and go to the big city in order to become someone special and achieve something really important. Another, in his quest for true goodness, joins a revolutionary cause or goes into the jungle to be a doctor for the underprivileged. There are countless different paths a human being can choose at this critical juncture, but they are all driven by the same need: the need to do that which he or she considers to be truly significant, meaningful, good, important, or moral. The drive for self-esteem, which compels a young child to crave his or her parents' approval, is the *same drive* that later in life might very well cause him or her to reject his parents' approval and creates a powerful determination to achieve any one of the things described above. In other words, the inborn drive for self-esteem is the same drive that compels us, as we mature intellectually and emotionally, to seek true meaning, purpose, significance, and morality. It is also important to realize that every society forms some type of value system as a *response* to this inner need.

There are only two possible sources for this inner inborn "moral imperative," this force that compels us to *crave self-esteem, be good, do what's right, and do what's important and significant even if it means (in certain cases) sacrificing our lives in the process.* We may be unclear or disagree on what the particulars are, but we are driven nonetheless.

This drive is either **(A)** totally and wholly the function of some chemical, molecular, or genetic source and can be understood and explained as a purely physiological process, or **(B)** it is a genuine and authentic metaphysical/spiritual need and drive that has been planted in us by God. As explained in Chapters 8 and 9, the only possible source for real (non-subjective) values and morality is from a transcendent God. That is to say, the reason I have an inner need to do what I believe is truly good is because there are, in absolute terms, values which *are* truly good. The reason I am prepared to die for certain values is because I have been hard-wired with awareness that there are spiritually existent values that actually are more important than my own life. (This awareness of values is only in a *general* sense. I do not think it is possible to support the assertion that the *details* of these values have been hard-wired into our souls. The details must be revealed to us by God.) God has programmed us to seek the absolute truths and values that will connect us to the infinite, eternal being of God himself (as was explained in Chapter 9). If option **(B)** is true, the "moral imperative" is ultimately the drive to seek a relationship with God. If option **(A)** is true, our moral sense is the result of a peculiar arrangement of genetic material in our DNA and the entirety of morality, values, and ethics is nothing more than a chemical illusion. It is crucial to understand that there are no other choices or options; the source is either purely material or it is from God. (Please remember that "something that no one has thought of yet" is in the category of *unreasonable* options.)

It is equally crucial to realize that the "inner moral imperative" is not to seek "*a*" value system, it is to seek "*the true*" value system. Richard Dawkins and I both agree that if our ideas and value systems are demonstrated to be false, we must abandon them. Why? We are not interested in an illusory system; only the truth will satisfy our inner need. This drive of which we have been speaking is also a drive *to seek the truth*; true meaning, true morality, true purpose, true value, true significance — a drive so strong we are prepared to die for it.

Self-esteem Is Not a Physical Need

It should be obvious at this point that the source of our drive for self-esteem cannot possibly be physical because self-esteem itself is not a physical need. It is quite possible for someone in perfect physical health to be reduced to suicidal despair by attacking his sense of self-esteem. I cannot acquire self-esteem by ingesting a liquid, gas, or solid.* I need *values and ideas* that I believe in. I need values and ideas that I believe are the truth. I need *accomplishments* that I believe are truly significant. None of these are quantifiable by any physical or material measurement. The following scenarios are presented to drive this point home. Although the issues raised below have already been touched upon in Chapter 7, they are worth repeating as we draw the final conclusions from our discussions and analysis of Morality, Spirituality, and Man's Search for Meaning.

1. A man's fiancé tells him she loves someone else and never wants to see him again. Is it the sound waves emanating from her mouth that devastate him? Of course this is ridiculous. The same thing would happen if he read the words in a letter. Do markings on a piece of paper cause the fluctuations in heartbeat, brainwaves, and respiration? What if he pleads with her to come back, and she says nothing, writes nothing, but makes an obscene gesture with a look of hatred and contempt on her face. Is it particular "body positions" that cause all the profound

* Psychiatric medications do not give a person self-esteem. At best they simply level the playing field by equalizing a chemical imbalance. The person must then go and create a life for himself just like all of us must do. If someone knew the chemical formula for self-esteem, he would patent it and instantly become the wealthiest person who ever lived.

reactions we have been talking about? There is no physical, chemical, or molecular process to explain what caused this devastating reaction in the first place.

2. Consider sincerely religious people. Their entire understanding of who and what they are, their relations with others, and sense of place in the universe all revolve around the concepts and values that stem from certain religious principles. Imagine one of these people has just completed a book that "proves" their religious beliefs are false. He or she would no doubt experience a certain level of trauma. Their sense of security has been shaken. Without question, monitors would have indicated profound changes in body chemistry, heartbeat, respiration, and brainwaves. What caused all this to happen? A purely physical, material, and clinical description would be roughly as follows:

   After staring at printed symbols on bound pieces of paper, profound changes occurred in the subject's respiration, heartbeat, etc. Possible cause: chemicals in ink or on paper entered through pores on subject's skin causing reaction. No evidence found to support this hypothesis. Possible cause: subject swallowed piece of paper causing reaction. No evidence found to support this hypothesis. Possible cause: Unknown gas emanating from bound papers... No evidence to support this hypothesis.
   Conclusion: Unable to detect any material link between bound papers with printed symbols and subject's internal bodily functions. Highly unlikely that bound papers containing printed symbols had any connection to detected changes. Can find no evidence of any physical cause whatsoever that explains fluctuations in monitored bodily functions.

3. The reason that an investigation would find no physical cause for the reaction is because there was none. Ideas, concepts, and information caused the reaction. Ideas and concepts exist in time but not space. They are not measurable, quantifiable, or detectable by any material means. I repeat the assertion I made earlier: the burden of proof is on those who claim that every step, every action, and reaction is definable in purely physical terms.

4. After twenty years on the job, a man's boss informs him that the company has to let him go. He is crushed. It beggars the imagination to attempt to explain the turmoil going on inside this individual solely within the paradigms of physics and chemistry.
5. Recall the case we discussed earlier in Chapter 7. At a family gathering someone reveals the terrible secret that you had hoped would stay hidden forever. You are engulfed by a burning, searing feeling of humiliation. Your face goes white with shame, your throat feels constricted, and you have difficulty breathing. What happened to make your body react so violently? *Nothing at all changed in the physical universe.* How is it possible to destroy someone's self-esteem with cruel words and insults or revealing a secret? Are there tiny pieces of chemicals that enter into the person along with the offending words? How is it possible to lift somebody with a genuine compliment or kind words? Are there mysterious molecular structures that float through the air and enter in the person's body through their ears along with heartfelt praise and acknowledgement? I repeat what I said earlier: I defy any scientist to find a material, chemical, or molecular basis to explain why vicious or compassionate words, scrawled symbols on a piece of paper, or physical gestures, can have such a profound effect on another human being.

We have shown that the profound human need for self-esteem cannot be explained or understood as a physical or material phenomenon. It clearly belongs to the spiritual dimension of our existence. The only possible source for this most potent of human drives is God himself. When we seek values and moral guidelines, we are not seeking artificial human constructs; we are hard-wired to seek the absolute truths that will connect us to the infinite, eternal being of God. Uncovering the exact details of these moral truths and values is an altogether different undertaking, but this does not change the conclusions that we have reached thus far.

In Closing

Throughout this book I have been very careful to bring evidence and confirmation for my assertions from the writings and statements of prominent non-believers themselves. As the title of this book makes clear, my premise is that the theist wields the decisive intellectual advantage and it is the non-theist who scrambles to affirm his position with non-falsifiable axioms and leaps of faith. In the person of Harvard geneticist Richard Lewontin, I have found a remarkably candid non-theist who confirms this premise:

> Our willingness to accept scientific claims that are against common sense is the key to an understanding of the real struggle between Science and the Supernatural. We take the side of science in spite of the patent absurdity of some of its constructs, in spite of its failure to fulfill many of its extravagant promises of health and life, in spite of the tolerance of the scientific community for unsubstantiated just-so stories, because we have a prior commitment, a commitment to naturalism.
>
> It is not that the methods and institutions of science somehow compel us to accept a material explanation of the phenomenal world, but, on the contrary, that we are forced by our *a priori* adherence to material causes to create an apparatus of investigation and a set of concepts that produce material explanation, no matter how counterintuitive, no matter how mystifying to the uninitiated. Moreover, that materialism is absolute, **for we cannot allow a Divine foot in the door**.[5]

"Credo Quia Absurdum," I believe [in materialism] because it is absurd. The razor sharp retort of Dr. David Berlinski to the statement above by Lewontin should probably be etched in stone at the National Academy of Sciences: "If one is obliged to accept absurdities for fear of a Divine Foot, imagine what prodigies of effort would be required were the rest of the Divine Torso found wedged at the door and with some justifiable irritation demanding to be let in?"[6]

More than forty years earlier, Nobel Prize-winning biologist Dr. George Wald, made the following declaration that was nearly identical in its implications regarding the leap of faith necessary to embrace a materialistic view of reality:

> There are only two possibilities as to how life arose. One is spontaneous generation arising to evolution; the other is a supernatural creative act of God. There is no third possibility...a supernatural creative act of God; I will not accept that philosophically because I do not want to believe in God. Therefore, I choose to believe in that which I know is scientifically impossible; spontaneous generation arising to evolution.[7]

Despite the fact that, as evidenced above, non-theists are clearly prepared to throw rational/scientific thinking out the window when it suits their agenda, nothing is more likely to elicit an aggrandized sense of superiority in the heart of the militant skeptic than when he has the opportunity to tauntingly wave the banner of science in the face of the "primitive" and "backward" believer. What exactly are the parameters of this "science" to which Lewontin and other card-carrying atheists have so faithfully pledged their allegiance and to which they are committed to defend, come what may?

- Science has nothing to tell us about how life began.
- Science has no explanation for consciousness and our unique sense of identity.
- Science has no material explanation for our miraculous ability to communicate through spoken and written language.
- While science can explain to us in great detail why we need to eat and breathe, it gives us no meaningful insight into the burning human needs for ultimate meaning, purpose, accomplishment, and abstract moral values (in other words, those things which make us uniquely human). It offers no explanation (other than, of course, speculative theories) why every other form of life on earth continues thriving and reproducing perfectly well without any of the above, **while human beings are unable to live without them**!

In short, science has nothing to tell us about who we are, where we came from, and where we are going. Then what *does* science tell us? The simple truth is (this may be a shock to some people) that for the most part science is preoccupied with describing and discovering *how the plumbing works and building sophisticated gadgets*. This is not in any way meant to imply that science is unimportant. Not only are we in awe of scientific accomplishments, we are all profoundly grateful for the benefits that have been bestowed upon mankind by advances in science and technology, particularly in the area of medicine. However, despite the fact that scientific progress has saved countless lives, science offers no insight at all into *why* a human life is so worth saving to begin with and what we are living *for* in the first place.

Talmudic sources describe the practices of an ancient pagan cult called Ba'al Pe'or. The adherents of this sect showed their devotion to their god by first defecating in front of his statue and then proceeding to engage in some of the more standard types of debauchery. Defecation as a form of worship might seem odd to us in the twenty-first century, but the obvious, even profound meaning behind this act was to unequivocally proclaim the glorification of, and an exclusive devotion to, the physical and material aspects of existence. If it were suggested that the above described scatology would be an appropriate expression of their own absolute commitment to naturalism and materialism, I imagine the reactions of Lewontin, Coyne, Harris, Dawkins, Ruse, Dennet, Pinker, et al., would be comically squeamish. Despite that, my guess is they would still feel right at home in a post-worship philosophical discussion with the naturalist/materialist "theologians" of the Ba'al Pe'or seminary. After all, they both have opted for a purely materialistic approach to life.

At this point it has become clear that not only is life itself the result of a Godly act of creation, but that our experiences of the rich spiritual dimensions of living — a soul that is separate from our physical bodies, our yearning for transcendent meaning and purpose, and our drive for moral truth — are not illusory but are real and actual. What I have tried to demonstrate is that unless we are prepared to commit our hearts and souls to a "comforting fiction" as Dawkins, Baggini, Harris, and others clearly have, it seems that we are seeking God (or avoiding him), one way or another, with almost every move we make as human beings.

Is that notion really so strange? Isn't this the simplest and most obvious explanation why the overwhelming majority of human beings who have lived on this planet have spent their lives steeped in some sort of religious/spiritual quest of one sort or another? Doesn't it make more sense that these powerful, relentless inner longings and desires that drive us — longings and desires so powerful that they are greater than our desire for life itself — actually *are* connected to a reality that is much greater than ourselves? When Sam Harris, despite himself, writes of a "sacred dimension to life," when Dawkins speaks of a deep, tempting urge to "worship" a maker or creator, when Sartre exclaims that his "whole being cries out for God," isn't the most straightforward conclusion that these feelings reflect the truth of who and what we are?

What are we to make of the sublime experiential reality that in the face of a child we catch a glimpse of a splendor so brilliant and glorious that even if "all the earth were parchment and all the oceans ink" we still would be at a loss to capture it in words? Is the explanation for this indescribable radiance and preciousness to be found in the realization that a child reflects the infinite glory of his or her Creator, or is the explanation to be found in its having been descended from an ancient fish with a "peculiar fin anatomy"? Does the beauty and wonder of a child and the depth and intensity of the love a parent feels for a child bear witness to the words of Genesis that the human being was created "in the image of God," or does it bear witness to the proposition that mankind rose out of a pond of scum and was created not in the image of God but *evolved* in the image of the bacterium and the cockroach? Is the truth that the significance of one baby transcends the entire universe with its billions and billions of galaxies or is the truth that a child is simply another "purposeless" arrangement of electrons and molecules, "bits of a star gone wrong," a "carbon based bag of water," and that in objective reality there is no difference between burning 100 pounds of coal or a 100 pounds of babies? What is most awesome and even frightening about this question is that *these are the only two choices*. Isn't the answer obvious?

For those prepared to embrace truth and follow wherever it leads, there can be no real doubt as to the actual existence of the One transcendent God who created life and Whom we seek with an intensity and a passion that is so faithfully expressed by the Psalmist:

As the hart cries out in thirst for the springs of water, so does my soul cry out in thirst...for the living God...[8]

End Notes

1 Abraham Twerski, MD, *Addictive Thinking: Understanding Self-Deception, 2nd ed.* (Center City, Minnesota: Hazelden, 1977), p. 24.

2 "How Can We Strengthen Children's Self-Esteem," www.Kidsource.com/kidsource/content2/strengthen_children_self.html

3 Ibid.

4 Ibid.

5 Richard Lewontin, "Billions and Billions of Demons," *New York Times Book Reviews*, January 9, 1997, http://www.drjbloom.com/Public%20files/Lewontin_Review.htm

6 Berlinski, *The Devil's Delusion*, p. 15.

7 George Wald, "The Origin of Life," *Scientific American*, August 1954, pp. 44–53.

8 Psalms 42:2–3.

Chapter 11

Closing Thoughts

What about Evil and Suffering in the World?

What about Evil Committed in the Name of Religion?

Does the Existence of Evil Religious People Imply the Non-Existence of God?

Living a Godly life consists of essentially two different stages. The first stage is attaining an intellectual and emotional clarity regarding the reality of God's existence; the second stage is seeking the true path that will lead an individual to a relationship with God. Unfortunately, after completing stage one, many people approach stage two employing less discretion and wisdom than they would in their search to buy a used car. How to go about finding a true path to a relationship with God is an entirely different undertaking than being certain in your mind and heart that God exists. It is also not the subject of this book. This is the reason why I did not make any lengthy attempts to address the issue of evil and suffering. Making sense out of evil and suffering is only necessary and meaningful as a function of building a relationship with an actually existent, infinite, transcendent God. In other words, if God exists, I need some kind of explanation for the terribly painful things that happen in life in order to properly relate to him. I will briefly elaborate.

The One God described in this book either exists or He does not. Either He created the universe or He did not. *The fact that the world is not exactly the way we would like it to be is not a factor in determining the truth of the existence or non-existence of God*. The truth of His existence or non-existence is independent of any subjective reactions we may have about the vicissitudes of life. To suggest otherwise is as intellectually absurd as the notion that my negative evaluation of a particular corporation implies the non-existence of the CEO. The existence of the CEO has no connection whatsoever to my personal feelings about the corporation or the way it is run. If God does exist, it is clear that the gulf between the capabilities of the human mind and that of a donkey is trivial compared to the infinite gulf between the human being and God. It would be foolish to expect a donkey to understand everything that we do, and it is even more foolish for us to expect to understand everything that God does. If God does exist, He will run His universe according to *His* will, purpose, and understanding, not ours. The reality of suffering and the existence of people who commit (what we consider to be) horrible crimes may very well interfere in our relationship with an existent God, but they do not determine, nor negate, His existence.

Not only would this be an intellectually indefensible position, but even on an emotional level this approach would be invalid. There is no such thing as a "standard" emotional/spiritual response to a traumatically painful experience. There are those who *lose* their faith as a result of tragedy, there are those who *maintain* or *grow in* their faith during an encounter with tragedy, and there are many who *find* their faith as a result of tragedy. In short, the question of suffering and evil in the world has to do with our *relationship* with God, not His existence. (On the other hand if God does not exist, the whole question ceases to have significance. The pain or pleasure of mankind turns out to be nothing more than the luck of the draw.)

Evil in the Name of Religion

Generally speaking, each of the major religions of the world — and I would imagine most of the minor ones also — claim to have a divinely revealed message for mankind. That is to say, each claims that its scripture does not just contain elements of truth and reality, but that only its scripture expresses and represents *the* truth and reality. Despite the noble intentions of politically correct people everywhere, it is self-evident that every religion cannot be the truth. This is apparent even to those with only a superficial understanding of the theologies of the world's major religions. The commonly bandied-about notion that "my beliefs are true for me, yours are true for you, and his are true for him," is in most cases a classic example of being so open-minded that "your brains fall out."

If, for example, five different faiths each contain dogmas and tenets that theologically and philosophically exclude any of the other four from being the absolute God-given truth, then, to put it bluntly, we are stuck with only two possibilities: (A) only one of them is *the* truth and the others are false, or (B) none of them are *the* truth and they are *all* false. One could propose that they each contain aspects of truth, but that doesn't solve this particular dilemma. You would now have to create a new religion containing all those little pieces of truth extracted from other sources, a religion that you would immediately proclaim as being the *real* truth. We would then of course be right back where we started from — with six religions instead of five. (It is important to understand that rejecting a religion's claim to truth does not preclude the practice of tolerance. Tolerance does not mean that I necessarily accept your ideas as being true or valid. It means that despite the fact that I disagree with you, I still respect your essential value as a human being and continue to relate to you with the sense of dignity and obligation due to someone created in the image of God.)

If you observe what you perceive to be evil behavior on the part of adherents of a particular faith or atrocities committed in the name of a particular religious belief, there are three possible explanations:

a. Their religion is false. What they claim as a divine message and a divinely commanded "moral code" is really nothing more than an imaginary human construct and their behavior reflects this falseness (it goes without saying that even if they behave in a way that one might find very pleasant, their religion *still* could be nothing more than a product of human imagination), or

b. Their religion or moral code is divinely revealed truth, but human beings through their own free will have chosen to do evil and distort what God intended, or

c. Their religion is divinely revealed truth, they are essentially living the way God wants them to live, and *you* are the one who has a distorted sense of how human beings are really supposed to live and behave. Another possibility is that you have an improper or incomplete understanding of why the adherents of this faith are behaving in a certain way, or you might be lacking accurate information. It is also possible that what you perceive as an atrocity is not really an atrocity at all. Take, for example, the destruction of Hiroshima by an atom bomb. Depending on one's perspective, it can be viewed as an absolutely justified, moral act that ultimately ended World War II and saved millions of lives, or it can be viewed as an unprecedented act of barbarism.

Does the existence of evil religious people imply the non-existence of God? Hardly. The fact that people may correctly and truthfully believe in God does not preclude the very real possibility that they have deluded themselves into thinking that their own particular code of behavior is divinely ordained. Human beings are also capable of choosing to do evil of their own free will despite what they outwardly proclaim to believe.

Where Does This Leave Us?

I would suggest that it is the responsibility of each individual to invest the time, effort, and energy necessary to discover the truth about the meaning, purpose, and direction of his or her own existence. Otherwise, we are all faced with three equally lamentable alternatives: (1) heedlessly following the path that our society has conditioned us to travel, (2) manufacturing our own comforting illusion, or (3) making ourselves "comfortably numb" and playing out our lives as aimless, rudderless pieces of driftwood following the path of least resistance.

I wish I had an easier answer for you, reader, but there is no escaping the simple fact that no one can do your work for you. Let truth become your passion and your obsession. Seek, investigate, question, contemplate, discover, evaluate, and ultimately make a decision as to what is true and what is false, what is real and what is fantasy. Our own minds, our own lives, and our own souls are all we have. To allow ourselves, due to laziness and distractions, to flounder in mindless insignificance and deception would be tragic beyond words. I urge the reader to consciously and resolutely choose to begin this (perhaps lengthy) journey and process without delay. The moment that choice is made, the *real* adventure of life begins.

Appendices

Appendix 1

The "Simple" Self-Replicating Molecule

It is worthwhile mentioning here, as a brief introduction to our subject, a famous experiment conducted by Dr. Stanley Miller (1930–2007), another leading researcher in the Origin of Life field. Miller burst onto the scene while still a graduate student at the University of Chicago. In 1953, under the guidance of his mentor, Nobel Prize-winning chemist Harold Urey, Miller set up a simple experiment that consisted of sending electric current through a chamber containing chemicals assumed to be present in Earth's early atmosphere. The results sent shockwaves through the scientific world. The experiment yielded a number of different chemical compounds, including amino acids, the "building blocks" of proteins and therefore of life. There was a triumphant feeling in the scientific world that if a simple electric current combined with young Earth chemicals could produce amino acids, the secret to an unguided, naturalistic Origin of Life was right around the corner. This experiment is routinely referred to in high school biology courses as a landmark moment in Origin of Life research. In truth, as chemist Gunter Wauchtershauser put it, the experiment was a "dead end." Noted Origin of Life expert, Dr. Paul Davies, explains:

> Making amino acids is what a physicist would call "thermodynami-

> cally downhill," which means it is a natural process that occurs automatically, like crystallization. But hooking the amino acids together into long chains to make proteins goes the other way. That is an "uphill" — a statistically more difficult or unlikely — process. Let me give you an analogy. It's a little bit like going for a walk in the countryside, coming across a pile of bricks and assuming that there will be a house around the corner. There is a big difference between a pile of bricks and a house.[1]

Dr. Graham Cairns-Smith, an organic chemist and molecular biologist at the University of Glasgow, offers his viewpoint on the Miller experiment:

> It is true that some of the simpler amino acids have been found in complex mixtures generated under conditions simulating those that might have been present on the primitive Earth...but all such "molecules of life" are always minority products and usually no more than trace products. Their detection often owes more to the skill of the experimenter than to any powerful tendency for the "molecules of life" to form... In sum, the ease of synthesis of the "molecules of life" has been greatly exaggerated.[2]

Sir Fred Hoyle also hammers away:

> To press the matter further, if there were a basic principle of matter which somehow drove organic systems toward life, its existence should easily be demonstrable in the laboratory. One could, for instance, take a swimming bath to represent the primordial soup. Fill it with any chemicals of a non-biological nature you please. Pump any gases over it, or through it, you please, and shine any kind of radiation on it that takes your fancy. Let the experiment proceed for a year and see how many of those 2,000 enzymes [needed for life] have appeared in the bath.
>
> I will give the answer, and so save the time and trouble and expense of actually doing the experiment. You would find nothing at all, except possibly for a tarry sludge composed of amino acids

> and other simple organic chemicals. How can I be so confident of this statement? Well if it were otherwise, the experiment would long since have been done and would be well known and famous throughout the world. The cost of it would be trivial compared to the cost of landing a man on the Moon... In short there is not a shred of evidence that life began in an organic soup here on earth.[3]

Dr. Robert Shapiro delivers the *coup de grace*:

> Let us sum up. The experiment formed by Miller yielded **tar** as its most abundant product... The very best Miller-Urey chemistry... does not take us very far along the path to a living organism. A mixture of simple chemicals, even one enriched in a few amino acids, no more resembles a bacterium than a small pile of real and nonsense words, each written on an individual scrap of paper, resembles the complete works of Shakespeare.[4]

We are now primed to explore the proposal that a simple self-replicating molecule was a precursor on the road to living cells. Could such a molecule — which is obviously far less complex than a bacterium — be generated by unguided, naturalistic forces? Richard Dawkins (a biologist, not a chemist) informs us that: "A molecule which makes copies of itself is not as difficult to imagine as it seems...the small building blocks were abundantly available in the soup surrounding the replicator."[5]

"*The small building blocks were abundantly available*"? As Dr. Paul Davies pointed out above, "building blocks" called bricks are everywhere, but there is a big difference between a pile of "building blocks" and a building. In fact, Dr. Robert Shapiro, professor emeritus of chemistry at NYU (self-declared agnostic, and opponent of Intelligent Design theory), commenting on the above cited passage, nonchalantly, condescendingly, and summarily dismisses Dawkins. After explaining that the simplest self-replicating RNA molecule would consist of about twenty nucleotides, or about 600 atoms bonded together in a strictly precise arrangement, he goes on to calculate the odds of such an event occurring:

> We badly need the point of view of the Skeptic once again... [Let's use the example of] the monkey at the typewriter. Let's call

> him Charlie the Chimp. Charlie...types out one line per second; completely at random...we want our monkey to type out "To be or not to be, that is the question," which has forty characters... The chances then become 45^{40} or about 10^{66} to 1. This number is ten million times greater than the number of trials maximally available for the random generation of a replicator on the early earth. There we have it. If the chances for getting the replicator at random from a prebiotic soup are less than that of striking "To be or not to be, that is the question," by chance on a typewriter, we had best forget it. The replicator would have about 600 atoms. The chances of Charlie typing a 600-letter message correctly are one in 10^{992}.[6]

At a conference of some of the world's leading scientists sponsored by *Edge*, held in August 2007, Shapiro was a little more candid about how he viewed Dawkins' understanding of the origin of life:

> Richard Dawkins wrote a wonderful book but the place where he absolutely blew it was in a section on the origin of life...he has no other recourse — he's not a chemist — than to invoke some improbable event...so his schoolboy howler is the section on origin of life.[7]

Dr. Shapiro also wrote the following in 1999, in anticipation of the creation in the laboratory of "self-sustained RNA evolving systems," i.e., the self-replicating molecules we have been mentioning:

> The media probably will announce it as the demonstration of a crucial step in the origin of life... The concept that the scientists are [actually] illustrating is one of Intelligent Design. No better term can be applied to a quest in which chemists...prepare a living system in the laboratory, using all the ingenuity and technical resources at their disposal.[8]

He goes on to say that watching top-level scientists manufacture these amazing pieces of molecular machinery in the laboratory evokes in him the same kind of awe as when he watches a skilled golfer play a difficult course at well under par. He concludes, "To imagine that related events could take place on their own appears as likely as the idea that the golf ball could play its own way around the course without the golfer."[9] I suggest to anyone who doubts this that they get hold of the published papers on these experiments and read the meticulously written and detailed lab protocols involved.

At a lecture sponsored by the Harvard Origins of Life Initiative in October of 2008, Shapiro had this to say about connecting prebiotic synthesis conducted in the laboratory with the theory that life evolved on Earth from naturally occurring RNA:

> While chemists have succeeded in making the molecules of life — or their components — in the lab out of simpler molecules...the tightly controlled processes in a chemistry lab can't be mistaken for what would have happened on the early Earth. "Any abiotically prepared replicator before the start of life is a fantasy."[10]

One of the essential problems is that nucleotides, the complex molecules that are the building blocks of RNA, have never been shown to form by any naturalistic process. They must be manufactured in the laboratory in a sequence of carefully controlled reactions via a discipline that scientists call "prebiotic [i.e., pre-life] synthesis."

Dr. G. C. Smith, in his book *Seven Clues to the Origin of Life*, points out the implausibility of naturally forming nucleotides. He explains that there are fourteen major chemical/molecular "hurdles" that must be overcome for nucleotides to form naturally in a primitive-Earth scenario. The only way that scientists know of to overcome these fourteen serious hurdles is by "organic synthesis," a precise manufacturing process that takes place in the laboratory. Each of these fourteen processes themselves consists of many separate laboratory operations involving "lifting, pouring, mixing, stirring, etc." Smith points out that while each separate operation by itself may not be that complicated, they must be carried out in a rigorously specific and exact sequence. When this manufacturing procedure is at all prolonged, *"it becomes absurd to imagine"* that such a process could have happened by chance on the primitive earth. Thus, Smith concludes, "simple amino acids are plausible prebiotic products, primed nucleotides are not." He then goes on to calculate the probability of nucleotides forming through a random process:

> But you may say, with all the time in the world, and so much world, the right combinations of circumstances would happen some time?... The answer is no: there was not enough time, and there was not enough world... It would be a safe oversimplification...to say that on average the fourteen hurdles that I referred to in the making of primed nucleotides would each take ten unit operations — that at least 140 little events would have to be appropriately sequenced. (If you doubt this, go and watch an organic chemist at work; look at all the things he actually does in bringing about what he would describe as "one step" in an organic synthesis.)...
> We can say that the odds against a successful unguided synthesis of a batch of primed nucleotides on the primitive Earth are similar to the odds against a six coming up every time with 140 throws of the dice... This is a huge number, represented approximately by a one followed by 109 zeroes. This is the sort of number of trials that you would have to make to have a reasonable chance of hitting on the one outcome that represents success. Throwing one dice once a second for the period of the Earth's history would only let you get through about 10^{15} trials, so you would need about 10^{94} dice. That is far more than the number of electrons in the observed Universe.*[11]

In the same vein, Dr. Robert Shapiro wryly notes: "Unfortunately, neither chemists nor laboratories were present on the early Earth to produce RNA," and adds, "Gerald F Joyce of the Scripps Research Institute and Leslie Orgel of the Salk Institute concluded that the spontaneous appearance of RNA chains on the lifeless Earth 'would have been a near miracle.'"[112]

At this point in our discussion it is probably worth noting what

---

* This is roughly equivalent to a snowball's chance in hell.

Nobel Prize-winning microbiologist Dr. Francois Jacob had to say on our subject: "It goes without saying that the emergence of this RNA and the transition to a DNA world implies an impressive number of stages, *each more improbable than the previous one.*" Dr. Gerald Joyce, a long-time collaborator of Leslie Orgel and one of the leading researchers in the area of self-replicating RNA molecules, essentially agrees with Shapiro (albeit writing in a more subdued tone) that nobody understands how the "RNA World" scenario could have developed under natural conditions:

> After contemplating the possibility of self-replicating ribozymes [RNA molecules] emerging from pools of random polynucleotides [building blocks of RNA] and recognizing the difficulties that must have been overcome for RNA replication to occur in a **realistic** prebiotic soup, the challenge must now be faced of constructing a **realistic** picture of the origin of the RNA World...it must be said that the details of this process remain obscure and are not likely to be known in the near future.[123]

Dr. Joyce's conclusions are echoed by Niles Lehman, evolutionary biochemist at Portland State University: "The odds of suddenly having a self-replicating RNA pop out of a prebiotic soup are vanishingly low."[134]In fact, in an article entitled "The Origins of the RNA World," Dr. Joyce explicitly points out that while the RNA World concept has helped guide scientific thinking and has focused experimental efforts, "this concept does not explain how life originated."

Finally, chemist Dr. Steve Benner, a world-recognized authority on Origin of Life research and chairman of the prestigious Origins of Life/Gordon Research Conference, has this to say about self-replicating RNA molecules:

> We have failed in any continuous way to provide a recipe that gets from the simple molecules that we know were present on early Earth to RNA...you [also] have a paradox that RNA enzymes, which are maybe catalytically active, are more likely to be active in the sense that destroys RNA rather than creates RNA... We are finding all sorts of problems in getting behavior that we find useful, let alone Darwinian out of this.[145]

After hearing world-class scientists in the Origin of Life field describe the possibility of a naturally forming "simple" RNA replicator as "vanishingly low," "absurd," "utterly perplexing," a "fantasy," a "near miracle," that a plausible explanation "remains unknown," that it's as likely to occur as a golf ball playing its way around an eighteen-hole course by itself, that as yet we have no "realistic" scenario for such an occurrence, that each of the many steps involved in such a process is "more improbable than the previous one," that to achieve such a feat you would need more throws of the dice than there are electrons in the universe, that in any case the entire area of research "does not explain how life originated," and that we have "failed in any continuous way" to solve the problem, one can only marvel at the chimerical approach to this subject taken by atheistic writer Rebecca Goldstein in her novel, *36 Arguments for the Existence of God — A Work of Fiction*:

> But the mathematician John Von Neumann proved in the 1950s that it is theoretically possible for a simple physical system to make exact copies of itself from surrounding materials. Since then, biologists and chemists have identified a number of naturally occurring molecules and crystals that can replicate in ways that can lead to natural selection...eventually leading to precursors of the replication system used by living organisms today.[16]

Since she is not a scientist, I think we can give her the benefit of the doubt and assume she was not engaging in a deliberate attempt to deceive readers with her gross misrepresentation of the facts; she is just disturbingly ill-informed. In short, the speculative suggestion of a "simple" self-replicating RNA molecule is another scientific dead end in the futile attempt to find a naturalistic origin of life.

End Notes

1 Dr. Paul Davies and Phillip Adams, "More Big Questions — In Search of Eden," http://www.abc.net.au/science/morebigquestions/stories/s540242.htm
2 Smith, *Seven Clues*, p. 44.
3 Hoyle, *The Intelligent Universe*, pp. 20–23.
4 Shapiro, *Origins*, pp. 105, 116.
5 Dawkins, *The Selfish Gene*, cited in Shapiro, *Origins*, p. 167.
6 Shapiro, *Origins*, pp. 168–9.
7 "Life: What a Concept!" published by *Edge*, 2008, p. 100, http://www.edge.org/documents/life/Life.pdf
8 Shapiro, *Planetary Dreams*, pp. 102–104.
9 Ibid.
10 "NYU chemist Robert Shapiro decries RNA-first possibility," *Harvard Gazette*, October 23, 2008.
11 Smith, *Seven Clues*, p. 46.
12 Dr. Robert Shapiro, "A Simpler Origin for Life," *Scientific American*, (Feb. 12, 2007), www.sciam.com/article.cfm?id=a-simpler-origin-for-life&page=1
13 Dr. Michael Robertson and Dr. Gerald Joyce, "The Origins of the RNA World," *Cold Spring Harbor Perspectives in Biology*, (Cold Spring Harbor Laboratory Press, April 28, 2010), p. 18.
14 Jeff Akst, "RNA World 2.0," *The Scientist*, March 1, 2014.
15 Susan Mazur, "Steve Benner: Origins Soufflé, Texas-Style," *Huffington Post*, Dec. 6, 2013.
16 Rebecca Goldstein, *36 Arguments for the Existence of God: A Work of Fiction* (New York, NY: Random House, 2010), p. 355.

Appendix 2

Taking the Next Step

(Neither of the lists below is in any way meant to be exhaustive)

1. For those who are not Jewish and would like to find out more about what Judaism has to say about seeking a path to God, the following are recommended:

   - *The Path of the Righteous Gentile* by Chaim Clorfene and Yaakov Rogalsky.
   - http://webpages.charter.net/chavurathbneinoach/index.html
   - http://www.aish.com/w/nj/

2. For Jews who are interested in finding out more about the Torah and Judaism, the following are recommended (in no particular order):

   - *Living Inspired; Worldmask*, by Rabbi Dr. Akiva Tatz, http://www.tatz.cc/bio.htm
   - *The Informed Soul: Introductory Encounters with Jewish Thought*, by Rabbi Dr. Dovid Gottlieb, http://www.dovidgottlieb.com
   - Rabbi David Aaron, http://www.isralight.org
   - http://www.shabbat.com/
   - http://www.partnersintorah.org/
   - http://www.gatewaysonline.com/index.asp
   - http://www.rabbiwein.com/

- http://lazerbrody.typepad.com/lazer_beams/
- http://www.aish.com/
- http://ohr.edu/
- http://jeffseidel.com/
- Books by Rabbi Shimon Apisdorf, http://www.allbookstores.com/author/Shimon_Apisdorf.html
- Rabbi Ari Kahn, http://www.rabbiarikahn.com/
- Books by Rabbi Aryeh Kaplan: *The Sabbath: Day of Eternity*, *Jerusalem: Eye of the Universe*, *The Infinite Light*, *If You Were God*, *Meditation and the Bible*, *Tefillin*, *Maimonides' Principles*, *A Call to the Infinite.*
- *Horeb* by Rabbi Samson Raphael Hirsch
- Rabbi Lawrence Kelemen, http://www.lawrencekelemen.com/
- Dr. Gerald Schroeder, *The Science of God: The Convergence of Scientific and Biblical Wisdom.*
- Rabbi Mordechai Becher, *Gateway to Judaism*
- *Why the Jews? The Reason for Anti-Semitism*, by Prager and Telushkin
- *The Living Torah* (Rabbi Aryeh Kaplan's translation of the Five Books of Moses)
- *The Stone Edition of the Five Books of Moses*, *The Stone Edition of Tanach* [*Tanach* = Pentateuch, Prophets, Writings](Artscroll Publishers)

About the Author

Moshe Averick received his ordination as an Orthodox Rabbi in 1980. He has taught theology and spirituality for nearly thirty years to Jews of all ages, ranging from high school to adult education, in the United States, Canada, and Israel. He is known for his singular ability to explain complex topics in clear, understandable language and — to borrow the description of one University of Chicago-trained philosopher of science — his "wicked" sense of humor. He supervised educational programs at UCLA and Northridge University on behalf of Yeshiva University of Los Angeles under the auspices of the renowned Simon Wiesenthal Center, and in Toronto, Ontario, he was among the founding faculty members of what is now one of North America's largest Jewish adult education centers, Aish Hatorah of Toronto.

Besides his career as an educator, Averick was a floor trader in the S&P pit at the Chicago Mercantile Exchange. He currently lives in Bet Shemesh, Israel and is the proud father of eight children and an ever-growing number of grandchildren.

About Mosaica Press

Mosaica Press is an independent publisher of Jewish books. Our authors include some of the most profound, interesting, and entertaining thinkers and writers in the Jewish community today. There is a great demand for high-quality Jewish works dealing with issues of the day — and Mosaica Press is helping fill that need. Our books are available around the world. Please visit us at www.mosaicapress.com or contact us at info@mosaicapress.com. We will be glad to hear from you.

MOSAICA PRESS